建筑立场系列丛书 No.61

U0735453

时间：空间记忆
Time: Memory in Space

汉英对照
（韩语版第377期）

韩国C3出版公社｜编

于风军 安雪花 孙探春 杜丹 焦明 徐雨晨｜译

大连理工大学出版社

俄罗斯智能学校 _CEBRA

丹麦建筑工作室CEBRA赢得了智能学校的设计竞赛。该项目位于俄罗斯伊尔库茨克，其获奖方案名为"新型学校"，也被称为智能学校。

智能学校项目的愿景是创造一所新型学校———一座"学校公园"，是融建筑和景观为一体的独特学习环境，同时也是当地社区中心。这个项目基于统一与扩大的理念而建，而这个理念将学校综合体、自然与现代社会连接在一起，从而建造了多样化的学习环境，并使其富有活力。该综合体设计为可渗透的环形结构。环形结构内设主要的功能，并且围绕着中央的室外空间设置。每幢建筑的功能都以直接与外部空间相连的形式向中心漫延，而外部的区域组成了用于体验式学习与协同教育的多样化竞技空间。中心校园内活跃的社交空间与其外围安静的自然农业景观形成了对比，因此这个校区可部分自给自足，并且同时也成为教学活动的一部分。整座建筑覆盖了高低不同的屋顶，这样这个综合体便有了辨识度，并且在错落的体量之间创造了一系列有顶的室外空间。这些区域创造出不同的过渡带，消除了内外空间之间的界限，形成了学习、体育活动、玩耍、社交和休闲的场所。建筑之间的空间也作为重点来突出。这些空间的设计有双重目的，既可以作为不同功能区域的入口，又可以将不同功能区域连为一体。这些有顶的公共空间结合了亲密与语境化的元素，将这一大型学校综合体建筑、中心校区与景观连接在一起。

这座智能学校教育综合体能容纳1040名学生和400多教职员工，并且可分别供专业团体和学校外的其他用户使用。整个综合体包括幼儿园、初中、高中以及文化、休闲和健康中心，并且向公众开放。学校非常重视多样性和包容性，所以学校里至少有15%的孤儿。因此，这里将会设置特殊的宿舍，以便孤儿可以和他们的抚养家庭住在一起。

Smart School in Russia

Danish architecture studio CEBRA has won a competition to design an education campus in Irkutsk, Russia, which the firm described as "a new kind of school", called the Smart School. The vision for the Smart School project is to create a new type of school – a "School Park" – that unites architecture and landscape into a unique learning environment and gathering point for the local community. The design of the project is based on a both unifying and expanding concept that bridges between school complex, nature and modern society to form a diverse, and activating learning environment. The complex is organized as a permeable ring of buildings, which contain the main functions and are grouped around a central outdoor space. Each building's function expands towards the center in the form of directly related outdoor area and together these areas form a diverse arena for experiential learning and pedagogical synergies. The very active and social spaces of this central campus are contrasted by a calmer landscape of natural and agricultural areas outside of the ring, which allow the school to be partially self-sufficient and are included in teaching activities.

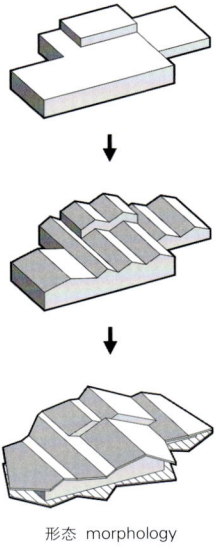

形态 morphology

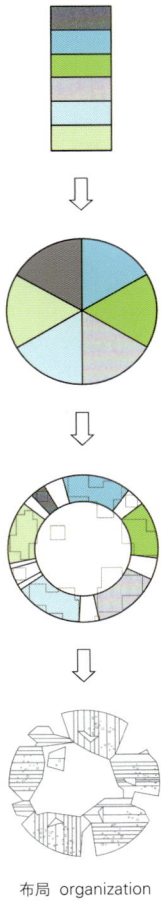

布局 organization

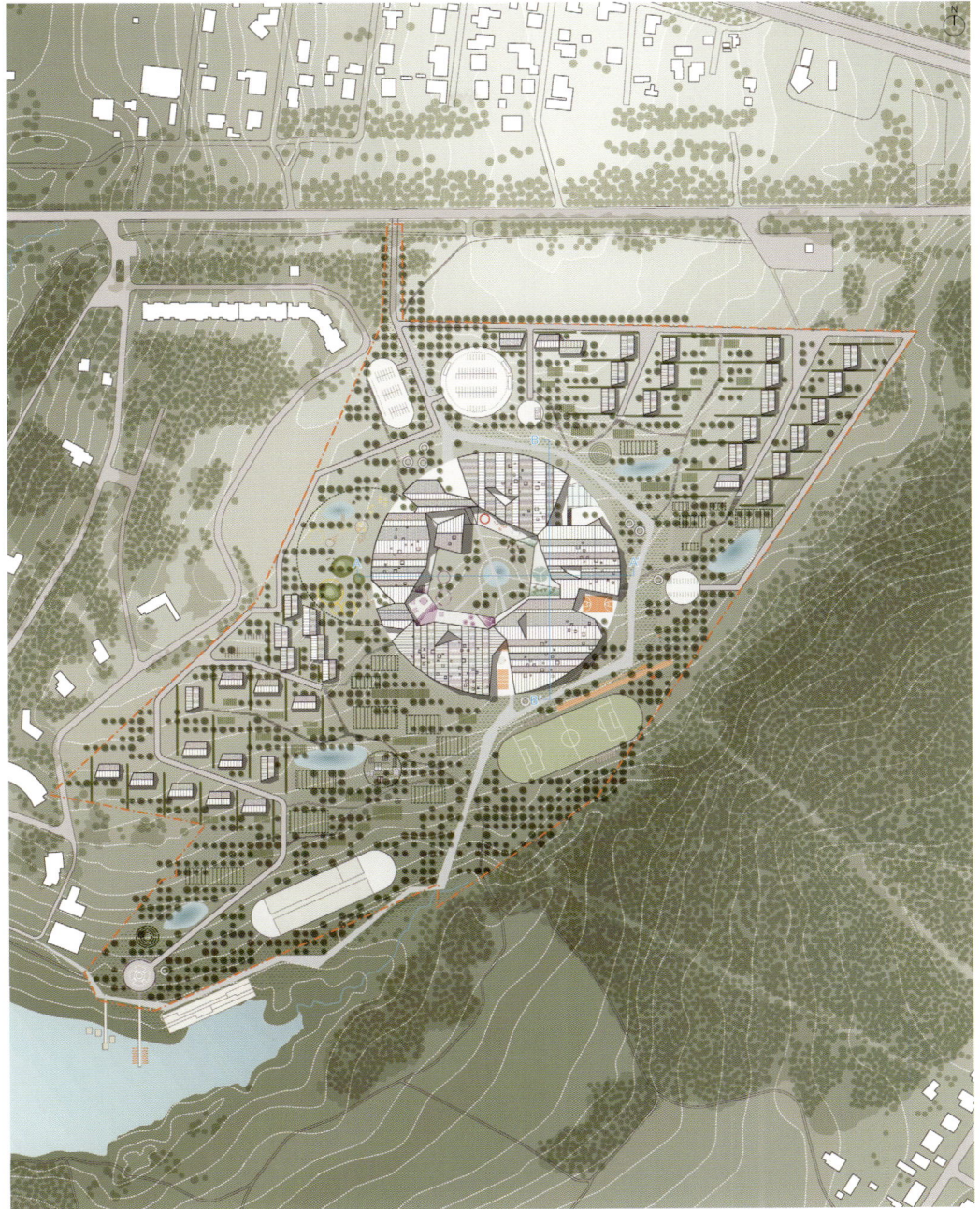

The buildings are connected by a large roof surface with distinctive eaves, which give the complex a recognizable identity and create a series of roofed outdoor areas between the offset volumes. These zones dissolve the boundary between the indoor and outdoor spaces by creating a diverse transitional zone for learning and physical activities, play, social interaction and relaxation. Special emphasis is put on the spaces in between the buildings. They are programmed for a double purpose in order to utilize their potential as both access points and links between different functions. These roofed common spaces combine the elements of intimacy and contextualization that tie the wider complex of school buildings, the central campus and the landscape together.

The Smart School's educational complex will have room for 1,040 students, a staff of over 400, divided into several professional groups, and outside users. The complex consists of a preschool, junior school and high school as well as cultural, leisure and health centers, which are accessible to the public. The school puts great emphasis on diversity and inclusion so that a minimum of 15% of the orphans are included. Therefore, the site will contain a special settlement, where the orphaned children live with their foster families.

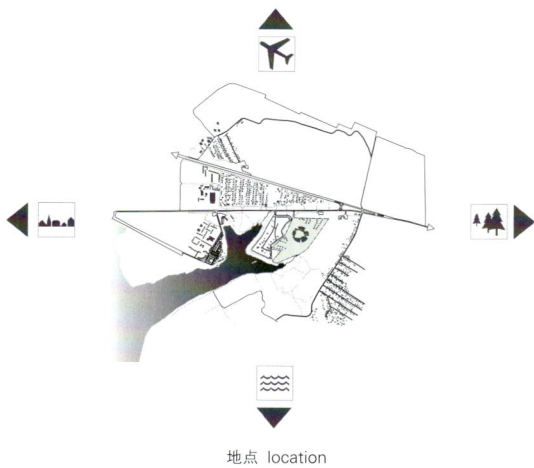

地点 location

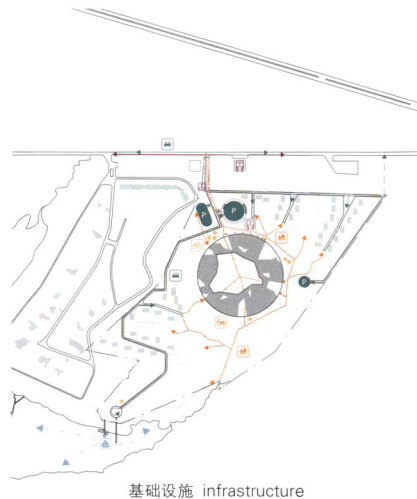

基础设施 infrastructure

parking
preschool

foster families (private)
farmland (semi private)
school yard
sport
public zone

区域 zones

1.寄养家庭的住宅 2.旅馆 3.初中 4.体育&健康中心 5.高中 6.温室&橘园 7.文化&休闲中心 8.主入口
1. houses for foster families 2. hotel 3. junior school 4. sport & health center 5. high school 6. green house & orangery 7. culture & leisure center 8. main entrance
东立面 east elevation

1.寄养家庭的住宅 2.旅馆 3.幼儿园 4.初中 5.体育&健康中心 6.高中
1. houses for foster families 2. hotel 3. preschool 4. junior school 5. sport & health center 6. high school
南立面 south elevation

1.幼儿园操场 2.幼儿园 3.行政楼 4.草地 5.高中 6.寄养家庭的住宅
1. preschool playground 2. preschool 3. administration building 4. meadow 5. high school 6. houses for foster families
A-A' 剖面图 section A-A'

1.寄养家庭的住宅 2.体育&健康中心 3.草地 4.文化&休闲中心
1. houses for foster families 2. sport & health center 3. meadow 4. culture & leisure center
B-B' 剖面图 section B-B'

隆德大学新论坛_Henning Larsen Architects

Henning Larsen建筑师事务所在隆德大学医学系的新医学广场的设计竞赛中胜出。作为与现有学院建筑之间的联系,在开放与透明的基础上,医学广场将会成为隆德大学员工、学生、研究人员和参观者新的聚集地。该建筑成为将小镇连接成一体的论坛。

室内的广场,即论坛,设有咖啡馆、餐厅和其他一些见面地点,可以随意进出,并与室外公共空间连接。开放的室内区域和室外区域融合成一个大的城市广场,在对外强调医学论坛功能的基础上,还为展览和非正式学习环境提供了空间。

广场顶部的建筑物呈45度角旋转。该旋转结构突出了建筑的身份与表达,并且使该建筑与Sölvegatan的周围环境相融。该建筑底部和顶部的对角旋转使该地区得到了最佳利用,并在建筑内外以壁龛和露台的形式提供了独特的休闲空间。

在这座新建筑的设计中,灵活性非常重要;现在的空间规划主要满足教学、研究与学习对空间的不同需求。设计包括一个简单的模块系统,该系统可以利用拉门等装置随时对空间布局加以修改。从长远来看,它也可以大幅度地改变建筑的空间布局,满足研究领域与教学形式不断变化的需求。

对该建筑而言,建筑材料的持久耐用和天然性非常重要。不同的材料用来表现医学院建筑的不同形式和功能,所以暖色系和冷色系材料被混合使用。钢材和木材是反复使用的材料,营造出融临床技术的精度和感官的温暖于一体的建筑氛围。

底楼的外立面和屋顶也都由钢材和木材装饰的表面覆盖,浸润日光,使人感觉置身室外,犹如在大树下乘凉。

New Forum at Lund University

Henning Larsen Architects has won a competition to design the new medical forum, an addition to the medical faculty at Lund University. With the open and transparent base as a link between the existing faculty buildings, the medical forum will become the new gathering place for employees, students, researchers and visitors at the Lund University. The architecture creates a uniting forum for the town.

The indoor plaza, the forum, will be an accessible area with cafe, restaurant, and meeting points, and continues the outdoor public areas. The open indoor and outdoor areas will fuse into one great urban forum, making room for exhibitions and an informal learning environment, while emphasizing the functions of medical forum for the outside world.

The building on top of the forum is rotated 45 degrees. The

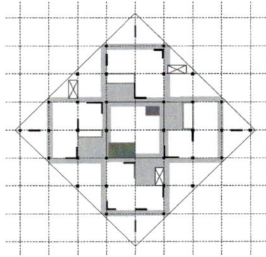

The floorplan offers great flexibility and can easily be adapted to future needs. The modular system creates spaces for teaching, educational and research purposes as well as workplaces.

灵活性 flexibility

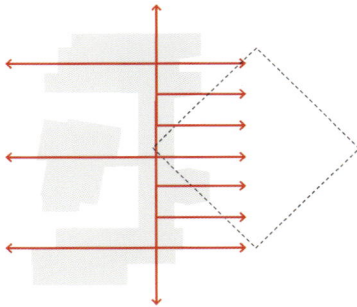

The new structure adapts to and connects with the physical surroundings.

社区 community

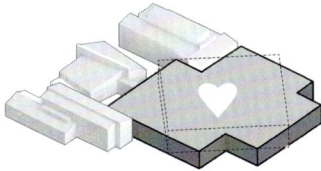

The additional building creates a public gathering point within the new structure.

社区 community

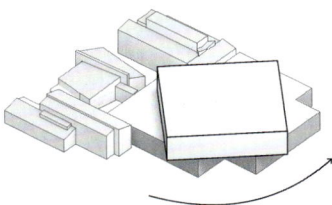

The twist of the top floors adds a unique identity to the new building which at the same time adapts to the existing road structure.

建筑特色 identity

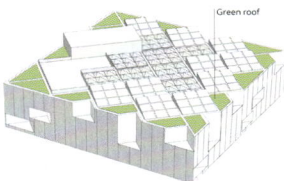

The flat roof of the main building is partially covered with a green roof that helps to increase biodiversity in the area and rainwater attenuation.

可持续性 sustainability

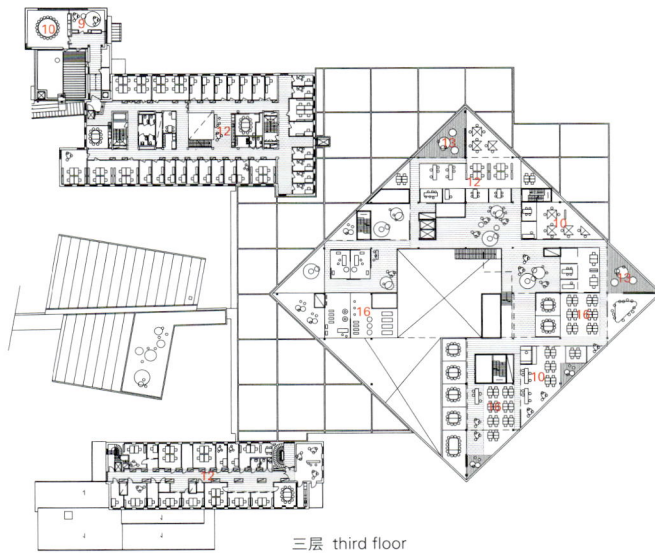

三层 third floor

1.咖啡厅
2.礼堂
3.商店
4.餐厅
5.厨房
6.多功能室
7.卫生间
8.实验室
9.休息室
10.小组讨论&会客室
11.接待处
12.办公室
13.露台
14.论坛广场
15.水塔广场
16.学习区

一层 first floor

1. cafe 2. auditorium 3. shop 4. restaurant 5. kitchen 6. multi-function room 7. toilet 8. laboratory
9. lounge 10. group & meeting rooms 11. reception 12. offices 13. terrace
14. forum square 15. water tower square 16. study area

twist strengthens the identity and expression of the building, and makes it blend in the context around Sölvegatan. The diagonal rotation of the bottom and top building utilizes the area optimally and provides unique recreational spaces in and outside the building in niches and terraces.

Flexibility has been essential in the design of the new building; the current space planning weighs the different needs that arise in an environment for teaching, research and study. The design consists of a simple modular system that can be modified from day to day, with the help of sliding doors and the likes, but also makes it possible to change the layout of the building more drastically in the long run, meeting the requirements of the changing needs for research areas and forms of teaching.

Durable and natural materials are significant for the building. Symbolizing the different forms and functions of the medical faculty, warm and cold materials are mixed. Steel and wood are recurrent materials, creating an atmosphere of the building that juxtaposes clinic precision and sensuous warmth. Facade and roof in the bottom building are also cladded with steel and wood, infiltrating the daylight and providing a feeling of being outside under the treetops.

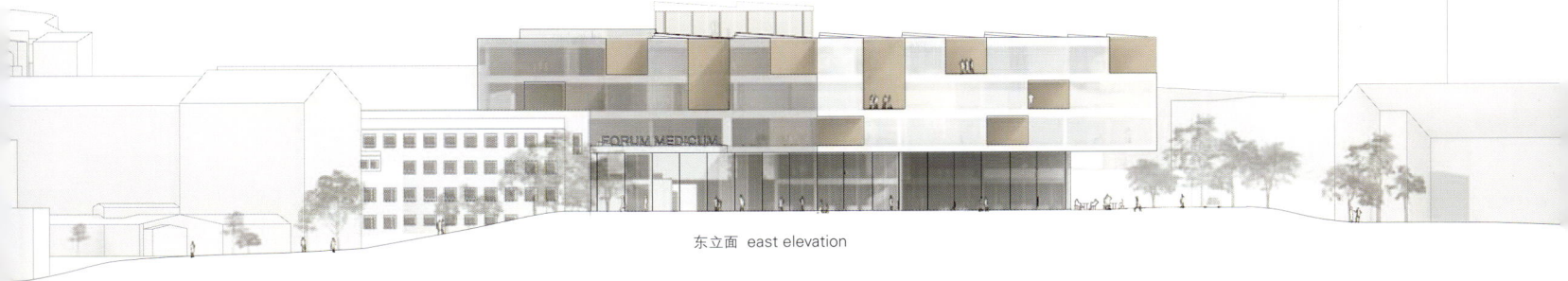

东立面 east elevation

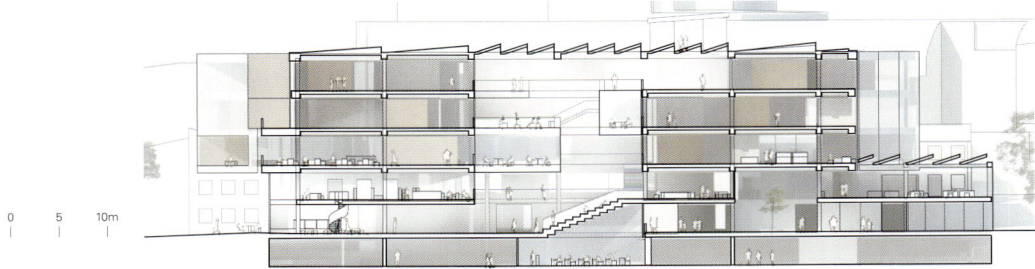

0 5 10m

A-A' 剖面图 section A-A'

布莱巴赫音乐厅 _ peter haimerl. architektur

该音乐厅是布莱巴赫市振兴新城市中心的核心项目。音乐厅紧邻新社区中心，是对由国家城市发展基金资助建设的新的村庄广场空间的有益补充。

音乐厅为单一混凝土体量，倾斜地置于村庄中心的坡地之上，因势利导，依地形地貌而建，立面装饰也采用了布莱巴赫市传统的建筑材料花岗岩碎石。

这栋整体倾斜的建筑的入口位于新的村庄广场，游客从入口进入，经过建筑下方楼梯到达地下门厅。门厅里设置有像衣物储存间、卫生间和酒吧这样的功能区域，同时还能引导顾客兴奋不已地绕过礼堂进入内部的音乐厅。

整个音乐大厅的设计遵循了声学原理，同时光从造型的缝隙中渗入，照亮环境。建筑主体由预制混凝土建造，只有高度复杂的构造模板才能实现这种建筑形式。音乐厅高耸而倾斜的表面基于声学规格建造，除了LED灯之外，还安装了应用于透光的缝隙后面和楼梯下面的消声器，以提供最理想的音响效果。大厅里的混凝土是未经处理的，其活泼生动的表面能够吸收中音音调。

建筑随着地形倾斜，内设走廊。固定在铁剑结构上的透明坐椅如同悬浮在透光缝隙之上。音乐厅的舞台使用了现代LED舞台技术，只为单一实际功能，而非多功能空间设计。

Concert Hall Blaibach

The concert hall represents the heart of the urban development to revitalize the new center of Blaibach. It is located next to the new community center and complements the space of a new village square that was realized with funds of the state urban development support.

The concert hall is a solitaire of concrete with an inclination above the slope in the village center following the topography and linking with its granite facade to the stone carver tradition of Bailbach.

The monolithic tilted building opens itself to the visitors at the new village square and guides them by a staircase to the foyer below the surface. The foyer provides not only the functional areas like wardrobe, sanitary rooms and bar, but also leads the visitor excitingly around the auditorium into the inner concert hall.

项目名称：Concert Hall Blaibach
地点：Blaibach, Germany
建筑师：peter haimerl. architektur
项目团队：Karl Landgraf, Ulrich Pape, Tomo Ichikawa,
Felicia Michael, Jutta Görlich, Martin Kloos
发起/操作：Thomas Bauer, Uta Hielscher
结构工程师：Thomas Beck, A.K.A. Ingenieure
供暖/空调工程师：Thomas Beck, A.K.A. Ingenieure
电气工程师：Planungsbüro Stefan Schmid
音效工程师：Müller-BBM
混凝土和室外混凝土模板制作：Fleischmann & Zankl
金属结构制作：Metallbau Gruber
室内混凝土模板制作：Gföllner, Fahrzeugbau und Containertechnik
主要赞助商：Euroboden Architekturkultur, Förderverein Konzerthaus Blaibach
瓷砖/淋浴设备设计：Duravit, Dornbracht / 纤维混凝土板设计：Eternit
油毡地面设计：Armstrong DLW / 音效系统设计：Akustik & Raum AG
甲方：Gemeinde Blaibach
用地面积：2,615m²
总建筑面积：575m²
有效楼层面积：500m²
设计时间：2013.1~9
施工时间：2013
竣工时间：2014
摄影师：©Edward Beierle (courtesy of the architect)

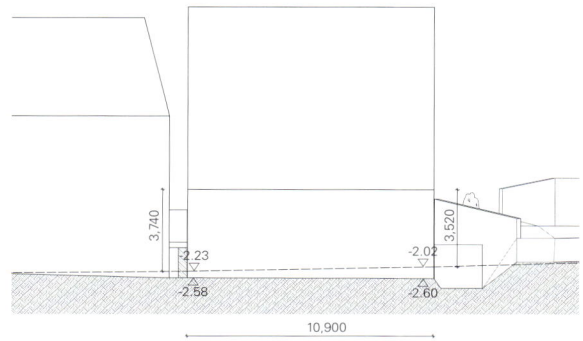

南立面 south elevation

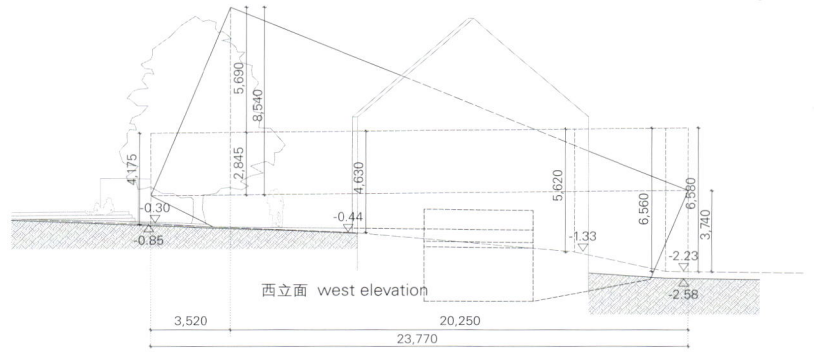

西立面 west elevation

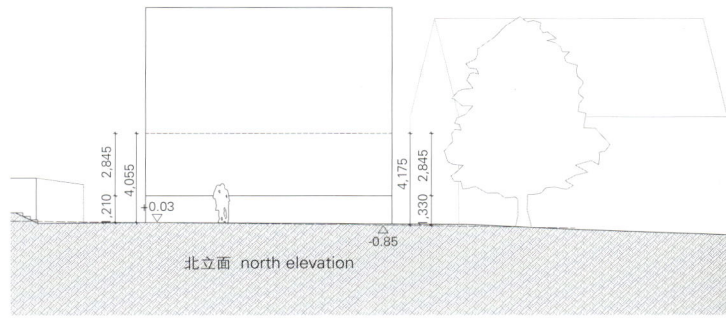

北立面 north elevation

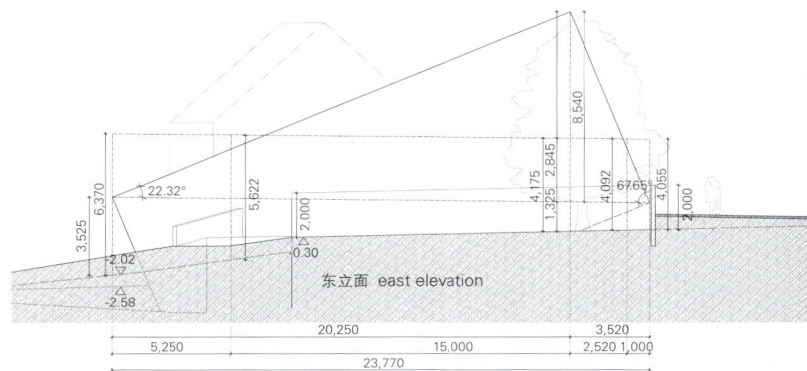

东立面 east elevation

The hall unfolds its acoustics within the seemingly light building, while the precise light slits illuminate the space. The building body is made of pre-cast concrete and only a highly intricate constructed formwork made the realization of the difficult form possible. The dominant tilted surfaces of the concert hall are based on acoustic specifications and include besides LED-lights also bass absorber behind the light slits as well as underneath the steps for optimal acoustics. The concrete in the hall is untreated. Its lively surfaces help to absorb the medium-height tones.

The inclination of the building – based on the increase of the slope – carries the gallery. The seemingly transparent seats, which are fixed on iron swords, appear to float above the light slits. The stage of the concert hall, which is only designed for its actual function not as a multifunctional room, is equipped with modern LED-stage technology.

1.卫生间 2.衣橱 3.技术间 4.舞台
1. toilet 2. wardrobe 3. technical room 4. stage
A–A' 剖面图 section A-A'

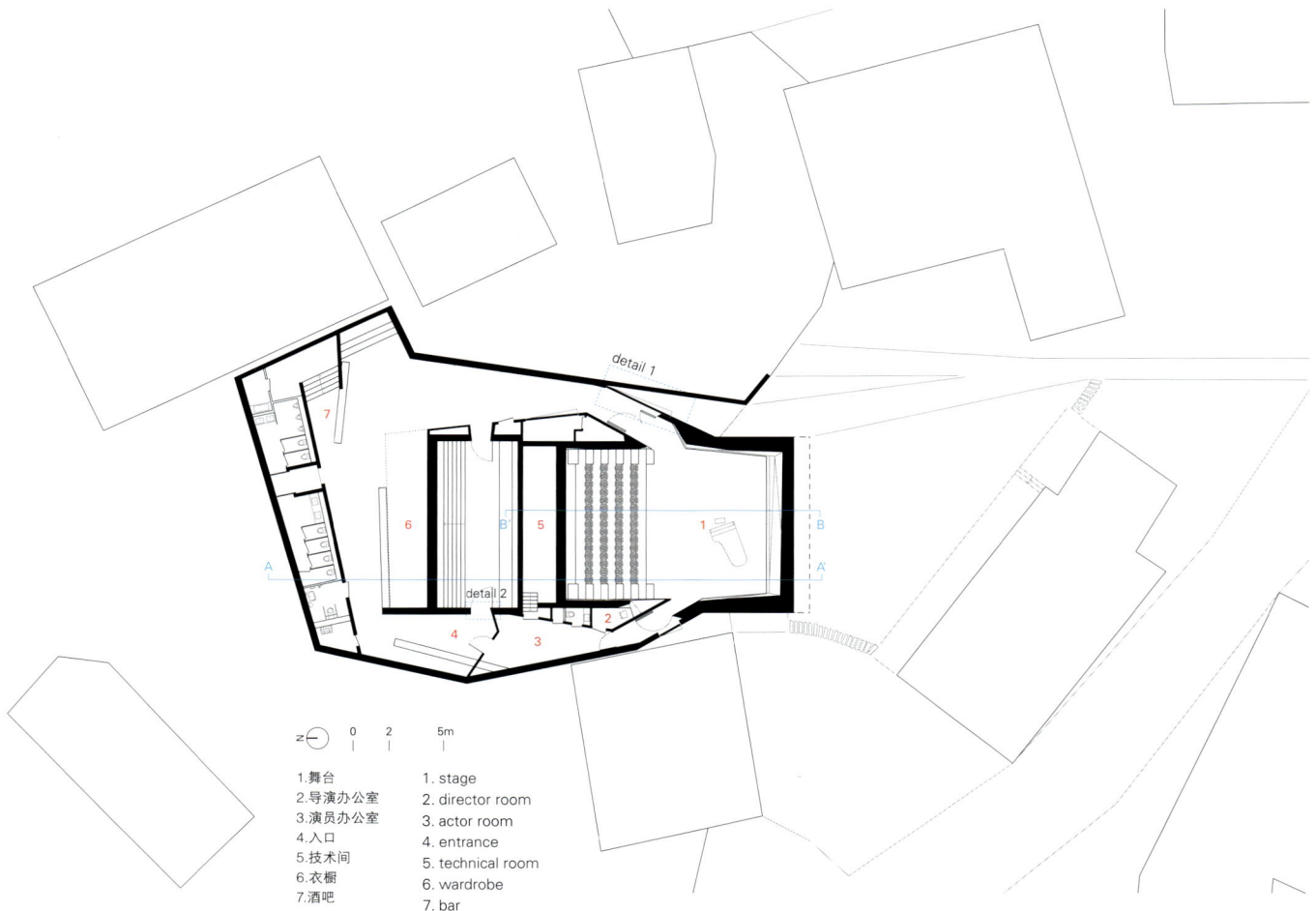

z 0 2 5m

1.舞台
2.导演办公室
3.演员办公室
4.入口
5.技术间
6.衣橱
7.酒吧

1. stage
2. director room
3. actor room
4. entrance
5. technical room
6. wardrobe
7. bar

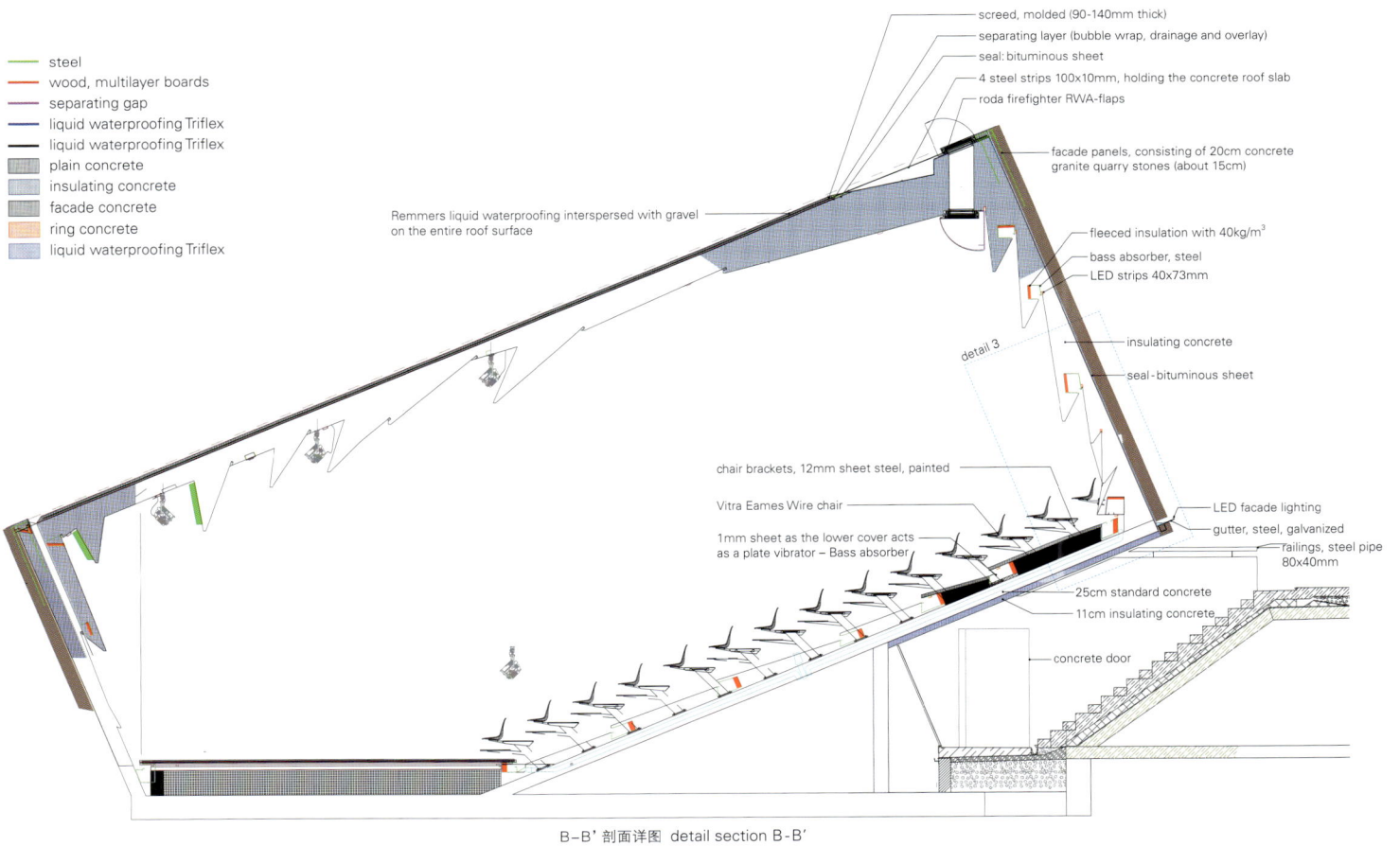

- steel
- wood, multilayer boards
- separating gap
- liquid waterproofing Triflex
- liquid waterproofing Triflex
- plain concrete
- insulating concrete
- facade concrete
- ring concrete
- liquid waterproofing Triflex

screed, molded (90-140mm thick)

separating layer (bubble wrap, drainage and overlay)

seal: bituminous sheet

4 steel strips 100x10mm, holding the concrete roof slab

roda firefighter RWA-flaps

facade panels, consisting of 20cm concrete granite quarry stones (about 15cm)

Remmers liquid waterproofing interspersed with gravel on the entire roof surface

fleeced insulation with 40kg/m³

bass absorber, steel

LED strips 40x73mm

insulating concrete

detail 3

seal-bituminous sheet

chair brackets, 12mm sheet steel, painted

Vitra Eames Wire chair

1mm sheet as the lower cover acts as a plate vibrator – Bass absorber

LED facade lighting

gutter, steel, galvanized

railings, steel pipe 80x40mm

25cm standard concrete

11cm insulating concrete

concrete door

B-B' 剖面详图 detail section B-B'

panic bar
DORMA PHB 3000

existing door frame

insulation

dictator final dampers
EDH 200

steel angle 250x50
on concrete wall

详图1 detail 1

filler

door handle flat steel 8x40mm,
V2A, and edged with 4-spindle
welded h=1,100mm

Styrodur filled with fabric
connects to wood panel

protect the rod with foil and
fill with expanding foam

详图2 detail 2

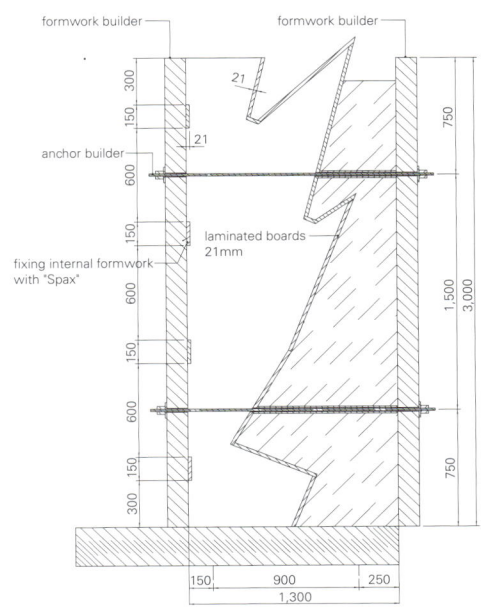

formwork builder

formwork builder

21

anchor builder

21

laminated boards
21mm

fixing internal formwork
with "Spax"

300

150

600

150

600

150

300

750

1,500

3,000

750

150

900

250

1,300

详图3 detail 3

四重奏之屋_Bluebottle

四重奏之屋是为庆祝弦乐四重奏及其悠久历史和与当代产生的共鸣而建,是一个享受视听盛宴的结构,其形状可以增强观众的视觉与听觉体验,并将表演空间与表演实践融为一体。每一场演出,四重奏之屋都只容纳最多50名观众;在一晚上的表演过程中,将为观众呈现几组风格迥异的曲目。

澳大利亚国家音乐学院是澳大利亚唯一的国家级的基于演出的音乐培训机构,把全国各地最优秀的青年音乐家聚集到一起,进行为期一年的高强度的学习、训练与表演。

该学院的迅速变化的艺术环境极具开放性,其项目得到了众多具有创新意识的合作伙伴的支持,包括获得国内外艺术家以及各流派的培训组织、表演组织和展出机构的支持。

作为富茨克雷社区艺术中心驻团艺术家,Bluebottle设计并建造了一处便携式空间。其内为艺术单元而设计的系统为了与四重奏之屋的模板相契合,已重新调整其设计和工程。

四重奏之屋将采用所设计的整套组件来建造。墙面板由胶合板建造,尺寸都是统一的,为1.2m×1.2m和1.2m×2.4m,用螺栓固定一起,形成所需的形状。然后,安装定制的面板。

建筑外立面表皮无缝链接,隐藏了出入口,可以自然而然地激发人们的好奇心。

Quartette House

Quartette House is a project that celebrates the string quartet, its rich history, and its contemporary resonance. It is a listening structure shaped to enhance the aural and visual experience of the audience and merge performance space with performance practice. For each performance Quartette House will hold up to 50 audience, and within the course of an evening several sessions of contrasting repertoire could be presented.

The Australian National Academy of Music is Australia's only national performance-based music training institution, bringing together the finest young musicians from around the country for an intensive year long program of study, training and performance.

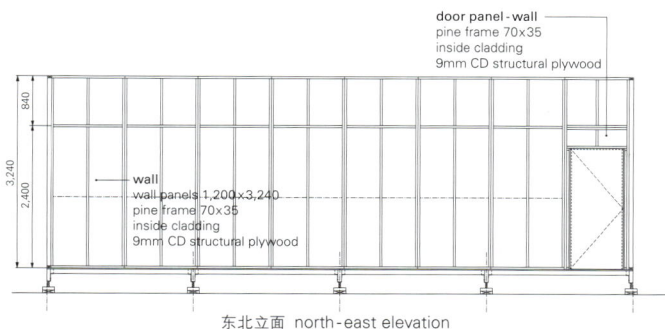

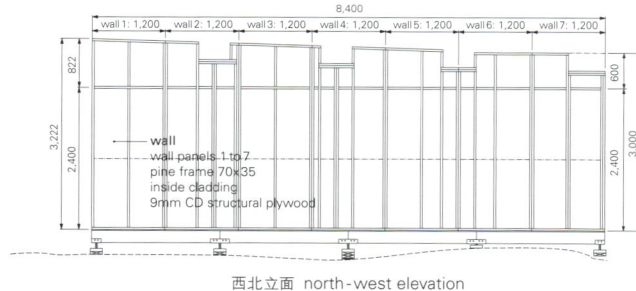

door panel - wall
pine frame 70x35
inside cladding
9mm CD structural plywood

840
3,240
2,400

wall
wall panels 1,200x3,240
pine frame 70x35
inside cladding
9mm CD structural plywood

东北立面 north - east elevation

8,400
wall 1: 1,200 wall 2: 1,200 wall 3: 1,200 wall 4: 1,200 wall 5: 1,200 wall 6: 1,200 wall 7: 1,200

822
3,222
2,400
600
3,000
2,400

wall
wall panels 1 to 7
pine frame 70x35
inside cladding
9mm CD structural plywood

西北立面 north - west elevation

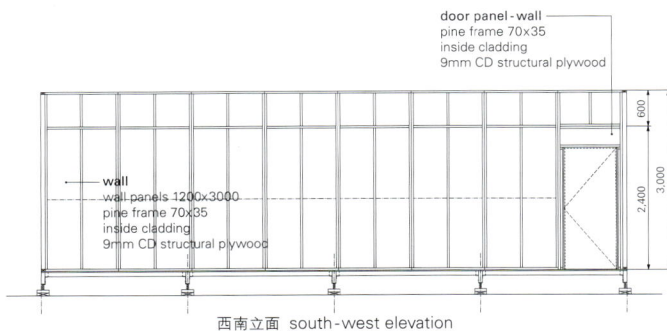

door panel - wall
pine frame 70x35
inside cladding
9mm CD structural plywood

wall
wall panels 1200x3000
pine frame 70x35
inside cladding
9mm CD structural plywood

600

3.000

2.400

西南立面 south-west elevation

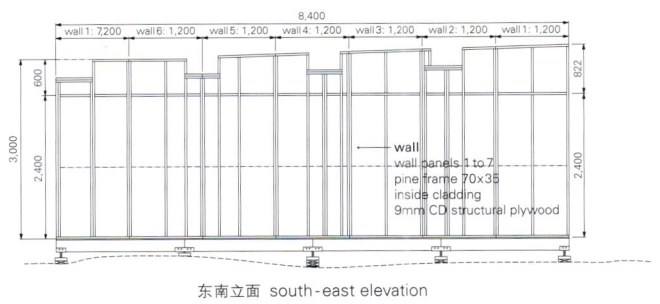

8,400

wall 1: 7,200 | wall 6: 1,200 | wall 5: 1,200 | wall 4: 1,200 | wall 3: 1,200 | wall 2: 1,200 | wall 1: 1,200

600

822

3.000

2.400

2.400

wall
wall panels 1 to 7
pine frame 70x35
inside cladding
9mm CD structural plywood

东南立面 south-east elevation

Open to the rich possibilities of the rapidly changing artistic environment, the Academy's programs are underpinned by a broad range of creative partnerships with national and international artists and training, performing and presenting organizations of all genres.

A portable space is designed and built by Bluebottle as artists in residence at Footscray Community Arts Center. The system designed and engineered for The Art Unit has been reconfigured to form the template for the Quartette House.

The Quartette House will be constructed using a kit of parts designed. Constructed out of plywood, the standard wall panels are 1.2 x 1.2m and 1.2 x 2.4m, and bolted together to form the desired shape. Custom panels will then be constructed.

The exterior will be clad with a seamless skin, allowing entrances and exits to be concealed, thus evoking a natural sense of curiosity.

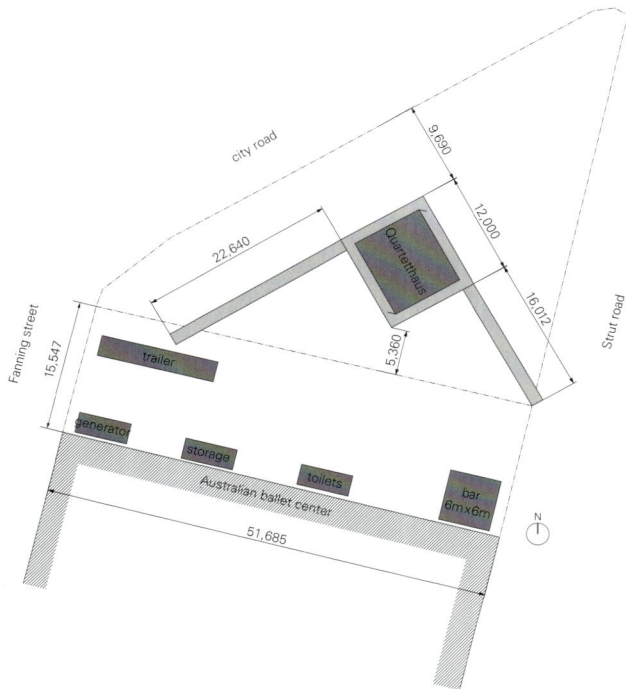

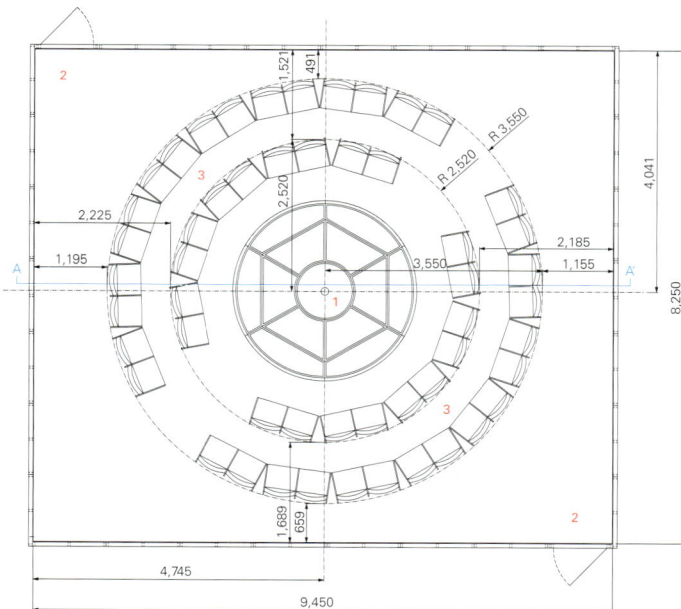

1.表演区 2.入口 3.坐席
1. performance area 2. entrance 3. seating
一层 first floor

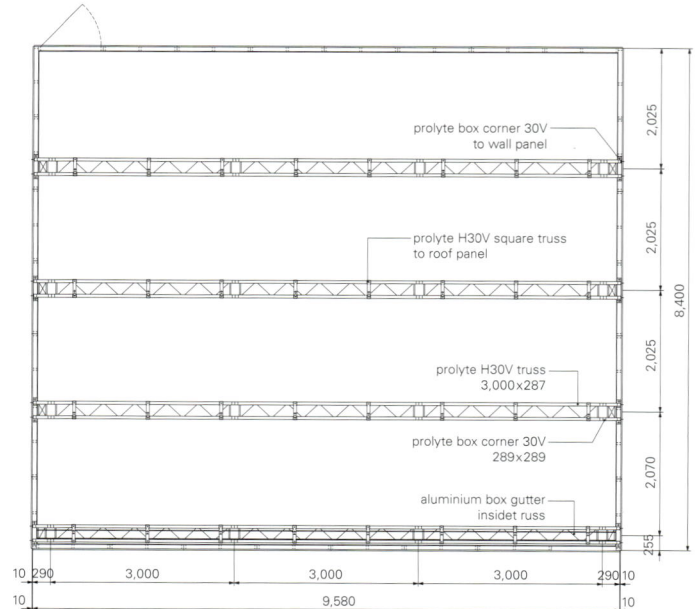

天花板_桁架 ceiling_truss

1.坐席 2.天花板嵌板 3.圆形幕墙 4.部分墙体嵌板组件
1. seating 2. ceiling panels 3. circular curtain 4. kit of parts wall panel
A-A' 剖面图 section A-A'

项目名称：Quartette House
地点：Portable venue-premier season Melbourne ACCA forecourt as part of Melbourne Festival
设计师：Bluebottle / 项目团队：Ben Cobham, Andrew Livingston_director design, Frog Peck_technical director, Paul Summers_general manager(site coordinator),
Marie-Pierre Dugrenier, Tom Rodgers_assistant designer(drafting), James Russell_technical co-ordinator(drafting)
合作者：ANAM (Australian National Academy of Music) / 总建筑面积：100 m² / 设计时间：2011 / 施工时间：2011 / 竣工时间：2011
摄影师：© John Gollings (courtesy of the architect) -p.22~23, p.25, ©Pia Johnson (courtesy of the architect) -p.24, p.26~27

Colonia Pictorilor画廊 _9 Optiune

　　巴亚马雷的艺术家殖民地开始于最初的1896年到1901年间每年夏季都发生的殖民聚居定居事件。1898年，由许多在巴亚马雷做长期或短期停留的艺术家形成的一个单一永久殖民区将这些临时定居点连为一个整体，这对建立这个被称做"Colonia Pictorilor"的永久性机构建筑起了决定性作用。如今，该建筑由许多展览空间与工作坊组成，为艺术家，即塑料艺术家联盟协会的成员提供了宽敞的创作空间。2012年设计并完成的展览馆是通向整个建筑综合体的人行通道，连接公共空间与创意空间，也是专门用来展览并出售出版物、广告宣传资料和推广艺术中心物品的地点。

　　这栋建筑由混凝土、金属和玻璃共同打造，营造出一处密闭而又开放的空间，是一幢令人向往并打开艺术世界大门的建筑。

　　该建筑极具创意，我们把这栋不大的建筑（仅27.5m²）按照一件雕塑来设计，里面到处弥漫着与众不同的创新意识。建筑外壳被赋予粗糙的石头纹理，将园区工作坊与展览馆周围的建筑融为一体。

　　建筑物正面摆放着雕塑家Gyori Csaba的金属质地的作品，这也代表着该建筑综合体名字的由来。

　　场地地形使我们能够将建筑建在与大街和人行道同样的水平面上，高出庭院，其开放的露台也朝向庭院。

Colonia Pictorilor Gallery

The Artist Colony in Baia Mare started from the initial summer colonizations which took place every year from 1896 till 1901. In 1898 these temporary settlements were joined by the individual permanent colonization of many artists who have spent in Baia Mare in shorter or longer periods of their artistic biography, thus being decisive in the establishment of a genuine permanent institutional structure known as "Colonia Pictorilor". Nowadays it is represented by multiple exhibition

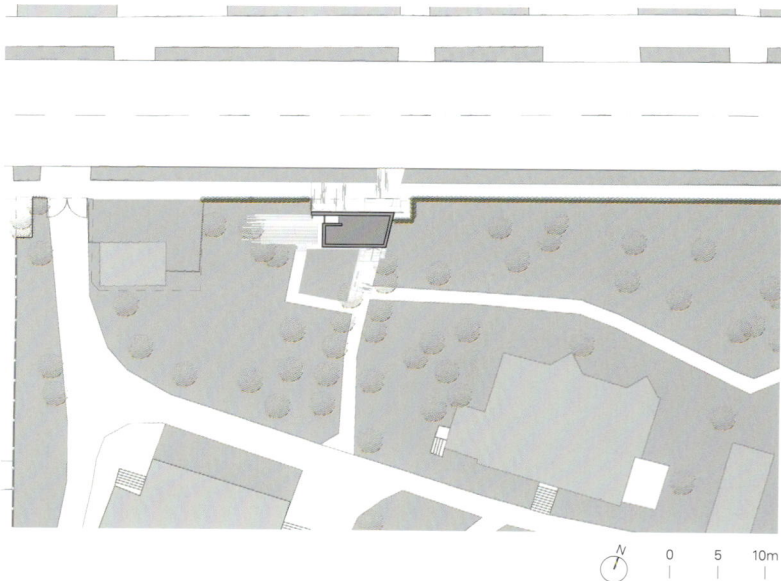

N　0　5　10m

spaces and workshops, a generous space offered to the artists, members of the Association of Plastic Artists Union.

The Pavilion designed and finished in 2012 marks the pedestrian access to the complex, being a junction point between the public space and the creative space, a spot dedicated to exhibition and sale of publications, advertising materials and objects promoting the artistic center.

The work was conceived as a composition in which concrete, metal and glass mingle, creating a closed and an open space

项目名称：Colonia Pictorilor Gallery
地点：Str. Victoriei nr. 21, 430141 Baia Mare, Maramures, Romania
建筑师：9 Optiune
有效楼层面积：27.5m² / 场地面积：148m²
设计时间：2012 / 竣工时间：2012
摄影师：
©Laura Teodora Ghinea(courtesy of the architects) - p.28, 30, 32
©Mihai Dan Mustea (courtesy of the architects) - p.29, 33

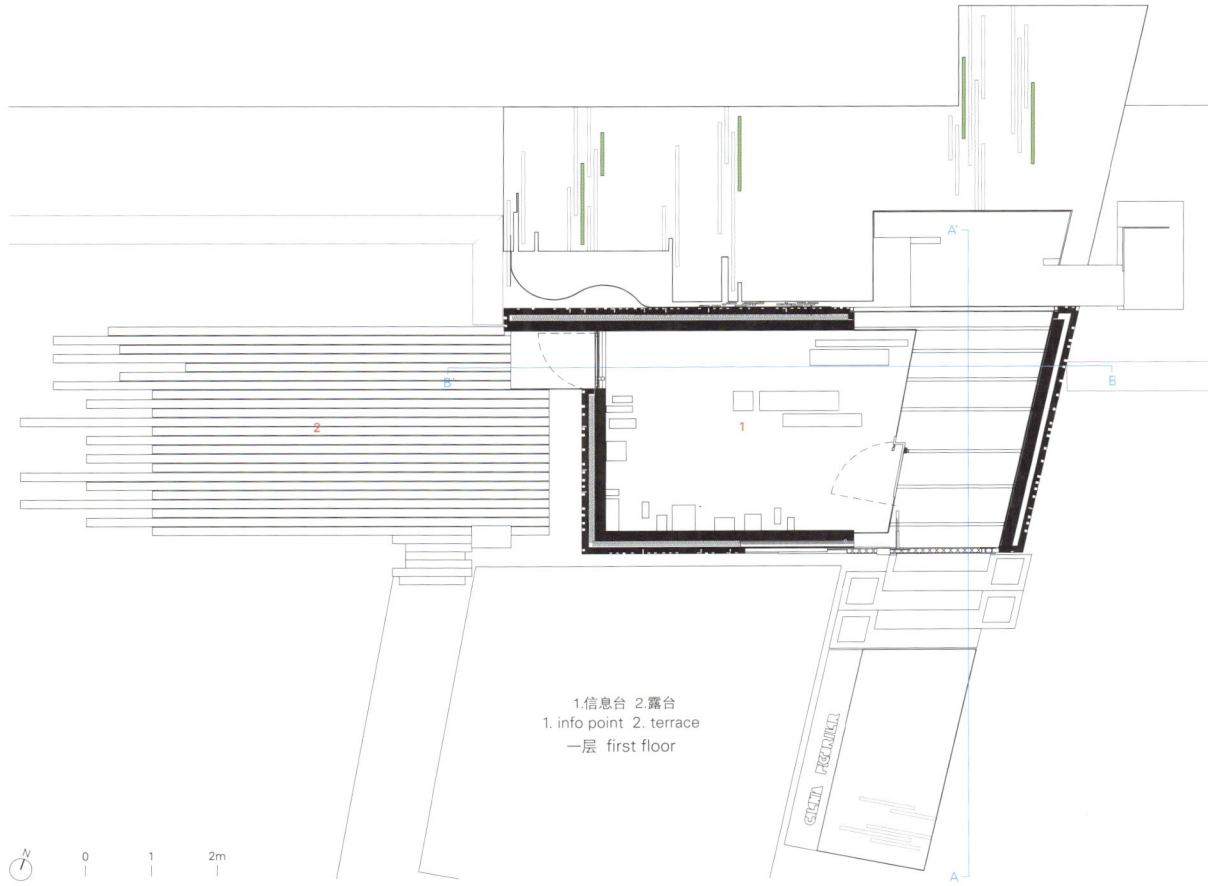

1.信息台 2.露台
1. info point 2. terrace
一层 first floor

N 0 1 2m

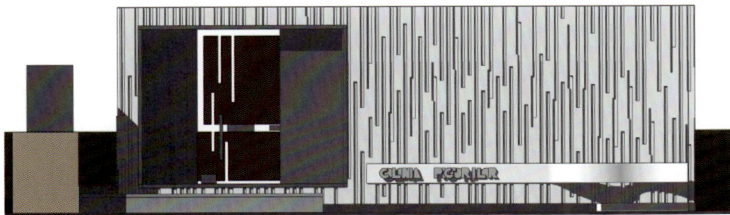

北立面 north elevation

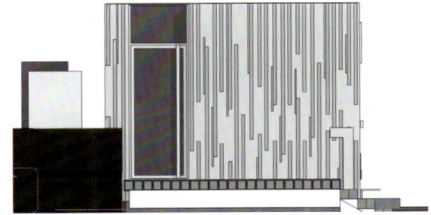

西立面 west elevation

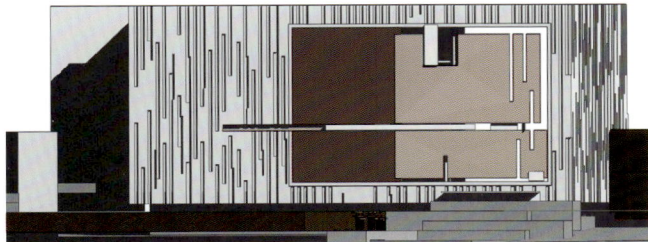

南立面 south elevation

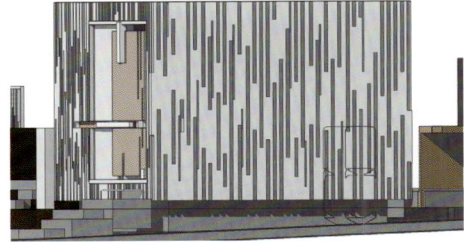

东立面 east elevation

at the same time, with an architecture that incites and opens the gates towards the world of art.

The building being an eminently creative space, we were determined to approach the small building (of only 27.5 sqm) as a sculptural object, in which innovation prevails over the common. The exterior shell has been interpreted to confer the rough stone texture that integrates the building in the park that surrounds the workshops and the exhibition pavilion.

On the front facade has been placed the metal work of the sculptor Gyori Csaba, which also presents the name of the complex.

The terrain shape has given us the possibility to create an object situated at the street and pedestrian level, that also floats towards the courtyard, to which the open terrace has also been orientated.

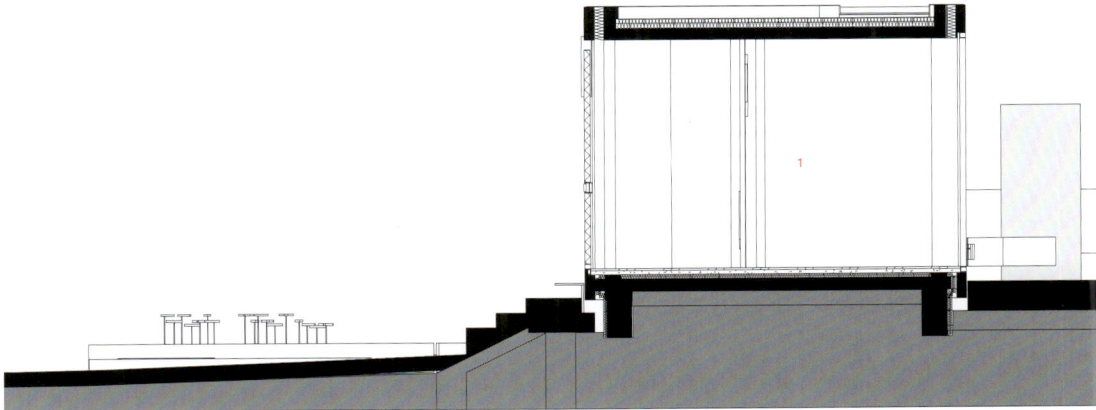

A-A' 剖面图 section A-A'

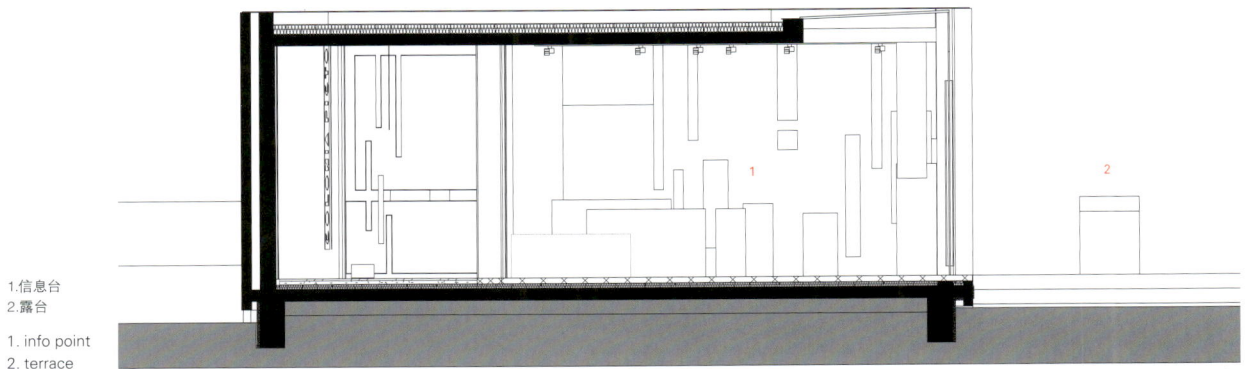

1.信息台
2.露台

1. info point
2. terrace

0   1   2m

B-B' 剖面图 section B-B'

时间: 空间记忆

Time: Memo

现在, 由于逐渐意识到资源浪费, 人们重新把历史建筑作为对话者来思考, 而不是作为障碍来拆除。建筑学所要解决的问题越来越多地与历史建筑的再利用、恢复和改造有关。随着时间的推移, 这些历史建筑渐渐地变成了遗产文物。因此, 设计者在设计的时候不仅要考虑空间因素而且还要考虑时间因素。在广义相对论被提出一百周年之际, 我们将会依据爱因斯坦的广义相对论来探究时空之间的联系, 这比任何其他理论都能富有诗意地阐释空间和时间之间那解不开的联系。

多亏了这位德国物理学家, 时间变成了一种物质, 一种人为可以操纵的有形物体。我们可以看到, 这种操纵体现在护身符的设计中。护身符通常是些很小的物体, 能够储存记忆, 能够扩大影响。对守卫护身符的人来说, 这种影响不仅与过去的故事有关, 而且能改变现在的环境。通过控制对过去事情的记忆, 人们能够创造一个更加美好的未来。

本篇文章中所呈现的这些项目通过记忆, 以各种各样的方法来操控时间, 但是, 每个项目在本质上都以"恋物"的定义为中心, 这极大地丰富了建筑物的空间复杂性。

More and more frequently architecture grapples with the reuse, restoration and adaptation of historic buildings that time has turned into heritage, as a mindset increasingly wary of resource waste forces their reconsideration as interlocutors rather than as obstacles to be demolished. Designers thus face work that contemplates time as well as space. On the centenary of its presentation, we will rely therefore on a line of inquiry that more than any other poetically explains the inextricable bond between space and time: Einstein's General Theory of Relativity.

Thanks to the German physicist, time has become matter, a tangible object to be manipulated – a manipulation we will see exemplified in the design of amulets, small objects that store memories, capable of extending an influence which, for the person who guards them, not only relates a past narrative, but alters the present environment. By manipulating the memory of past events one can build a richer future.

The projects presented here manipulate time, through memory, in various ways, but the essence of each centers on the definition of spatial fetishes which greatly enrich the spatial complexity of the building.

ry in Space

第7条法则：只要不得已，战斗就将继续。

1996年，恰克·帕拉尼克出版了他的第一部小说《搏击俱乐部》。小说展现给读者的是一个扭曲的世界，被主角所经历的事情在时间上的不断变形所扭曲：这是一个由现实中许多无情的眼泪组成的故事，故事的主角有着惊人的双重人格，他既是一个生活百无聊赖、患有失眠症的保险顾问——故事中没有名字的故事讲述者，又是一个有点不受约束的无政府主义者——泰勒·德顿。

因此，《搏击俱乐部》第七条法则成为这本书本身的宣言：只要不得已，小说就操纵时间，根据主角行为错综复杂的状态打乱情节，并且能够修改帕拉尼克所描述的宇宙中相同的物理现实；这就是摧毁我们所认识的那个世界，去实现另外一种美，这种美打破了常规，甚至打破了一些最基本的规则，比如时间从来不能倒流。

时间+空间（现在100年了）

2015年是物理学家阿尔伯特·爱因斯坦的《广义相对论》出版一百周年。有了这个重要发现，才有可能想象帕拉尼克笔下所描述的那个世界。在帕拉尼克笔下，这个世界貌似完全真实存在，人们可以在时间和空间纬度之间灵活穿越。

事实上，爱因斯坦的解释已经彻底改变了我们对于宇宙物理性质的了解，在三维空间的基础上增加了第四个维度——时间，在几何上四维空间合为一体，不可分割；也就是说，我们在世界上的运动，在其发生的环境中，占据空间的物体和事件会更改时间本身，即时间感知或时间流动。因此，时间并不存在线性流动：时间流动的进程取决于观察者，观察者的速度以及观察者的重量。这个德国物理学家用一个非常重要的、了不起的公式来总结这一理论，$E=mc^2$。在这个公式中，宇宙

Rule Number 7: The fights go on as long as they have to.

When in 1996 Chuck Palahniuk published his first novel, *Fight Club*, readers were projected into a view of a world distorted by continuous temporal deformation of the events the protagonist experiences: It is a story comprised of brutal tears in reality, in which the protagonist himself lives with an amazing duality of personalities, incarnated in the bored and sleepless insurance consultant, the unnamed narrator, and in a kind of wild anarchist, Tyler Durden.

The seventh rule of Fight Club, then, becomes a manifesto of the book itself: The narrative manipulates time, as long as it has to, to rip apart the plot, as conducted by the complexity of the behaviors of the protagonist, and it is able to modify the same physical reality of the universe described by Palahniuk; it is the destruction of the world as we know it, to reach another beauty, one that undermines the current rules, even the most basic ones, such as that time must flow in one direction.

Space + Time (Now 100 years old)

2015 is the centenary of the publication of the *General Theory of Relativity* by physicist Albert Einstein. With this essential discovery it has become possible to imagine the world described by Palahniuk as completely plausible in its agile movement between the dimensions of space and time.

Einstein's explanation, in fact, has fundamentally changed our approach to the physics of the universe, adding the fourth dimension of time to the three dimensions of space, all geometrically and inseparably combined: Our movement in the world, then, takes place in an environment where time itself, its perception, its flow, is modified by objects and events that occupy space. There is not, therefore, a linear passage of time: Its progress depends on the observer, on his speed, and on his mass.

In what has since become the epic formula that sums up the theory of the German physicist, $E=mc^2$, the energy of the universe, the mass of bodies, and the speed of light linked together and together necessitate change in not just the

23.2大宅，温哥华，加拿大
23.2 House in Vancouver, Canada

的能量，物体的质量以及光速相互紧密联系在一起，不仅迫使时间感知做出改变而且还迫使时间流发生改变。

那么，从某种意义上说，有了爱因斯坦，人们才揭示出操纵时间是我们环境设计的一部分。揭示出对过去记忆的修改或对未来的想象等同于眼下的工作，这样，时间就是一种物质，可以变成第四空间维度。

记忆的设计，或者相关的护身符

如在《搏击俱乐部》里人们可以操纵时间那样，在距恰克·帕拉尼克家乡南部几千公里之外的地方，从当地某些风俗中，人们也可以看到对时间的操纵；例如，在古代的印加文明地区，修改时间以一个objets trouvés系统的标准定义为基础，经过详细的设计，objets trouvés变成护身符，可以保护人的生命以及家庭和部落的安危。

由于护身符有记忆储存的功能，因此它就是过去的一个浓缩点，

集中反映了过去：被长期猎杀的动物或者是被赶走的危险捕食者；被击败的敌人；由神灵在过去的几个世纪中打造的稀有宝石；有着神奇特效、能够拯救部落成员生命的植物。这些物体的来源各不相同，多种多样，但是每一个护身符都通过自身反映了过去发生的事情，甚至拥有对佩戴者施加仁慈影响的能力，或者具有承载那些过去发生的事情并使其代代相传的能力，因为事实上，恰巧护身符都非常易于携带。

这样，护身符可以来自于动物、事物和人，并且成功地采用动物、事物和人所具有的品质和力量。这样，操控时间就变成有形的活动，时间本身通过纪念和记忆成为空间。

但是事实上，护身符带给人们的回忆并不是一种口头讲述，不会向人们讲述部落里伙伴们的故事，而是物质的，实际上存在的，是日常生活中随身携带的东西。从宗教方面来说，它是护身符，是一个神圣的东西，其神圣性从佩戴者每天的姿势动作和所经历的事情上得以体

perception, but the actual flow of time.

In a sense, then, thanks to Einstein, it has been revealed that the manipulation of time is part of the design of our environment and that the modification of memory of the past or the imagination of the future is equivalent to the work in the present. Time, then, is matter, which can be transformed into a fourth spatial dimension.

The Design of Memories, or Concerning Amulets

The manipulation of time, as shown in *Fight Club*, can also be observed in certain customs practiced a few thousand kilometers south of the home of Chuck Palahniuk: In the ancient territories of the Inca civilization, the modification of time is based, for example, on the definition of a system of *objets trouvés*, to be transformed into *amulets*, through detailed design, to ensure the protection of lives, of family, of tribe. The amulet is a point of concentration of the past, thanks to its function as a store of memories: the long-hunted animal or dangerous predator pushed away; the defeated enemy; the

rare stone, produced in the past centuries by a deity; the plant with miraculous virtues that saved a member of the tribe. These are objects of quite diverse origins, but each channels through its being the events of the past, even possessing the ability to extend a benevolent influence upon the wearer or to carry those events along, conveniently processed so as to be, in fact, transportable.

In this way amulets manage to adopt the qualities and abilities of the animals, things, and people from which the talismans derive. The manipulation of time, then, becomes tangible, as a matter of space itself, through remembrance and memory.

But the amulet is not, in fact, an oral narrative memory that addresses the fellow of the tribe: it is, rather, material, physically real, and something to take along in daily life. It is, religiously, an amulet, an object whose sacredness radiates in everyday gestures and events experienced by the person wearing it. These objets trouvés, reduced to portable talismans, exemplify a work of reconstruction of memory, carried out through

高山谷仓公寓，博西尼，斯洛文尼亚
Alpine Barn Apartment in Bohinj, Slovenia

现，这些objets trouvés成为便携护身符，在它们身上，通过把时间作为额外的空间坐标来体现对记忆的重构。在护身符身上实现了爱因斯坦关于空间和时间连续性的假设，但从某种意义上来讲，比起科学性更具有美感。

建筑——护身符，内／外

这里所提到的项目仅仅代表一种可能的出路，体现在一个美国作家、一个德国物理学家和一个亚马逊的部落的作品中。

多亏了爱因斯坦，我们得到了启示，或者说我们被赐予了一次机会，就是必须要考虑到建筑师的工作；也就是说，在设计中不能局限于规范的三维空间坐标，而是应该融入时间坐标。在帕拉尼克的作品中，我们可以看到一个将时间变形的世界。在这个透视空间的关键元素中，时间变得零散、破裂、迷乱而混杂。

最后，亚马逊部落告诉我们，时间是日常生活中的真实物质，因此，时间也可以改变；时间的改变是通过一种宗教仪式来完成的，（通过特定的、正能量的事件）将过去无形的事件变成一个重新设计的物体。这个物体就是护身符。护身符的设计表现出其神奇的属性。利用其属性，现在的日常生活和未来可能的日常生活可以改变。

在本章要介绍的设计项目中，护身符的神圣感是构成其建筑形象的基础：由奥马尔·阿贝尔设计的位于加拿大温哥华的23.2大宅，其设计出发点就是用几个大型的有着百年历史的木横梁来定义房屋屋顶的支撑结构。这些历史悠久的木横梁是从一系列烧毁的仓库中拯救出来的。

整个房屋设计利用了地面的坡度，其建筑形式也与这些木横梁的形状相适应——有一些横梁特别大。整个设计风格明显受到强烈的时间存在感的影响，时间改变了建筑。这些木横梁在设计者的语言中就是神圣的文物，其关键身份就是建筑的护身符，它们向人们讲述了谁创

the use of time as an additional spatial coordinate. In them is realized, in a sense more artistic than scientific, the Einstein's postulate concerning the continuous nature of space and time.

Architecture – Amulets, Inside/Outside

The projects presented here merely represent a possible path, here exemplified in the works of an American writer, a German physicist and an Amazonian tribe.

Thanks to Einstein, we may observe a revelation – an opportunity – that the work of an architect must somehow take into account: namely, that the coordinates within which one designs cannot be confined to the canonical three dimensions of space, but should expand to embrace the time coordinate. In Palahniuk's work we observe a world that transforms time – ripped, torn, confused – in key elements of spatial scenography.

The Amazonian tribes, finally, show us that time is a real matter of everyday life, and, as such, may be modifiable; it acts through a religious process that transforms the past (via particular positive events) from an intangible event into a re-designed object that is able to alter, through its properties, the everyday life of the present and of the possible future thanks to its miraculous properties, expressed through the amulet's design.

And it is the sacredness of the object that underlies the image of the projects presented here: In the 23.2 House designed by Omer Arbel, in Vancouver, Canada, the starting point of the project is defined by the discovery of several large centennial wooden beams, redeemed from a series of burned warehouses and used to define the supporting roof structure of the house.

Exploiting the slope of the land and adapting the building's forms to the geometric shape of the beams – some quite huge – the house assumes features clearly influenced by a strong presence of time, which changes its very being. The wooden beams, sacred artifacts in the words of the designer, acquire a key identity as architectural amulets which

照片提供: ©Quangtran

FA住宅, 大叻, 越南
FA House in Dalat, Vietnam

造了它们, 谁建造了它们所支撑的建筑, 以及它们最终是怎样回到现在, 现在又将怎样把这座房屋带到未来的。

因此说, 建筑物就变成了某种箱子, 保存着关于神圣物体的记忆, 既守护着记忆, 反过来又被记忆本身所操纵。

FA住宅就是这样, 但其设计方法更加激进, 该住宅由越南大叻的tho. A工作室设计。

这座旧房子经过宗教式风格的处理, 如同经过亚马逊部落护身符处理的建筑一样: 经过清洗, 除去了旧石灰墙面, 外表刷了一层新漆。新的外表让人们想起了建筑起初的样子。然后, 从屋顶开始的处理使其变得更加简单, 只有几根支撑横梁依稀可见。这样, 房子就变形成为一个神圣的护身符, 经过重新设计后, 成为保存记忆和突出其主要特征的物体。从这一建筑核心理念开始, 通过使用有形的建筑元素, 尤其是具有时间概念的元素, 这座住宅给人们带来一种新的空间感。

为了实现这种全新的空间感, 设计者通过一个抽象设计来定义这个护身符周围前所未有的环境: 这一抽象设计就是一个几何层, 既装饰了房屋又创造出许多模糊空间 (为建筑物的现貌带来了新的氛围)。半透明的蜂窝状聚碳酸酯涂层界定了一个非常不真实的三维空间, 其复杂而又模糊的环境让人质疑其本身是身在室内还是室外。

本章要分析的其他项目设计也是基于记忆和时间操纵这一主题, 也是把过去看作是一种崇拜对象, 但是, 在这些设计中, 新旧之间的关系在几何图形中是颠倒的。

在OFIS建筑事务所设计的斯洛文尼亚的高山谷仓公寓, Schemata建筑师事务所设计的日本鸠谷某住宅, 还有Baumhauer设计的位于瑞士塔拉斯普小镇的弗罗林斯公寓中, 贯穿于这三个项目的理念都可归结于把过去当作房屋的保护因素。建筑不再是要保存下来的护身符, 而是要具有庇护生活在其中的家庭的日常事务和未来事务的价值。

embody the story of who produced them and the buildings they supported, and of how they have finally returned to the present another temporality that will now bring the house into the future.

The building thus becomes a kind of chest that preserves the memory of the sacred objects, a body which guards, but which in turn is manipulated by, the memory itself.

Such is what happens in FA House, designed by atelier tho. A in Dalat, Vietnam, but via a more radical approach.

Here an old house undergoes an almost religious treatment, similar to that undergone by the Amazonian talismans: cleaned, stripped of the old plaster, covered with a new coat that recalls the first "version" of the building and then purified from the roof, with only a few support beams remaining visible. It thus transforms itself into a sacred amulet, an object redesigned to preserve memory and enhance its main features. From this architectural centre, via elements not only physical but also, and above all, temporal, there radiates a sense of the new space of the residential building.

To achieve this new sense, the designer defines an unprecedented setting around the amulet through an abstract operation: a geometrical layer which upholsters the house and creates blurry spaces that become the new ambience of the architectural present. A semi-transparent coating made of alveolar polycarbonate delimits an almost unreal three-dimensional enclosure in which outside and inside are called into question by the complexity of the blurry environments. The other analyzed projects work on the theme of memory and temporal manipulation, again through their treatment of the past as a kind of fetish, but in this case with the relationship between old and new geometrically reversed.

In Alpine Barn Apartment, designed by OFIS Arkitekti, in Slovenia, in the House in Hatogaya, Japan, by Schemata Architects and in the Florins Residence, by Baumhauer in Tarasp, Switzerland, the concept that links the three projects is ascribable to treatment of the past as a protective element of the house. No longer a talisman to be preserved, it acquires the value of covering the daily and future events of the families who will live here.

照片提供：©Kenta Hasegawa

鸠谷某住宅，埼玉县，日本
House in Hatogaya in Saitama, Japan

建筑护身符就是一个"肚子"，经过翻新和清洁，又变得漂漂亮亮，拿得出手，它是光滑的记忆，尽管被过去的伤疤所代表的苦难所净化，记忆里面仍包含着过去的伤疤。因此，在旧建筑物的内部建造新建筑，壳中有壳，一切仿佛就是出于对旧建筑物本能的尊重。

在由OFIS和Baumhauer设计的项目中，小型建筑作品似乎扎根在这些历史容器中，就像俄罗斯套娃一样，在更小的体块中还包含一些新的特征。在上述第一个项目中，这些小的建筑体都局限在古代建筑（护身符）之中，在第二个项目中，小的建筑体向外突出，好像在拼命展示自己的身份与存在。

在Schemata建筑师事务所设计的房屋中，由于其历史记忆没有特别的空间质量，再加上预算有限，因此其设计过程要更加巧妙，更加不易察觉，也更加困难。在这个项目中，原有的房屋由内而外被打扫得干干净净，外立面只成为一处布景。因此，新的设计策略需要对窗框的设计进行精雕细琢，如同对待艺术作品，在没有任何变化的外表和复杂的内部之间进行巧妙过渡。新设计主要围绕现有空置地面面积的不足来设计，来释放过于集中的房间布局。

与过去进行对话在建筑活动中日益重要，因它在操控记忆中发现了新的机遇：这个被称为偶像、法宝、或者护身符的元素，逐渐为周围环境注入了一种神秘感，使新建筑更加的丰富多彩，更加复杂多样，就像现在充满了丰富的过去。

The architectural amulet is a belly, refurbished, cleaned, again made presentable. It is a sleek memory which contains the scars of the past, though purified by the suffering of which they are token. And then, inside, the new is built, shell within shell, as if out of instinctive respect for the old.
In the projects by OFIS and by Baumhauer small works of architecture seem to stabilize within these historic containers, enclosing new features in smaller bodies, in the manner of a Matryoshka. In the first case they remain enclosed within limits set by the ancient architecture, the amulet; in the second they protrude outward, as if screaming out their own identity. In the house designed by Schemata Architects the process is more subtle, less visible and more difficult because of a historical memory with no particular spatial quality and thanks to a limited budget. In this case the existing house is cleaned from the inside, remaining externally a piece of scenography. The new strategy, therefore, necessitates the detailed work of framing the windows, treating them as works of art, of strategic passages between an unchanged exterior and a complex interior. The new activities revolve around a lack, the emptying of part of the existing floor, to release the excessive concentration of the rooms.
Working with the past, an increasingly important activity in architecture, finds in the manipulation of memory a new opportunity: The object – call it fetish, talisman, or amulet – instills in the ambient a feeling of sacredness, making the new richer and more complex, as if to conjure forth into the present the richness of the old existence. Diego Terna

高山谷仓公寓

OFIS Arhitekti

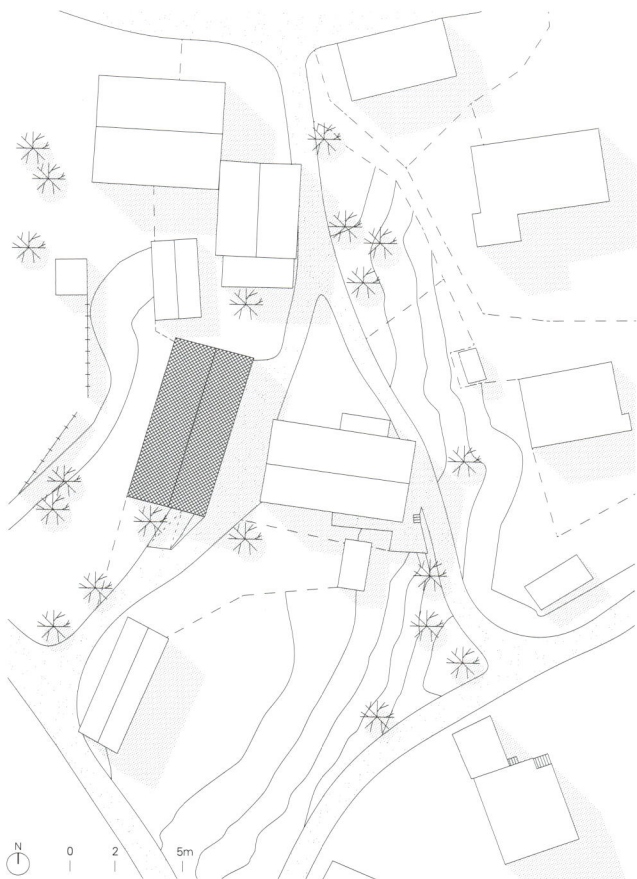

斯洛文尼亚农村的传统地标包括许多不同类型的农舍、干草架和谷仓。但是，不幸的是，许多已经不再使用了，因此大部分都有所破损，并且没有得到维护，通常只能摧毁并重建为一般的住宅。只有继承祖先留下的传统，斯洛文尼亚乡土建筑才能得以幸存下来。这些乡土建筑不仅是国家的象征，也是斯洛文尼亚农村生活不可或缺的一部分。

旧谷仓的重生

设计师的设计理念是把旧谷仓改造成一个阁楼公寓，保留原先的建筑外观，创造一个新的木质内壳，使两者形成鲜明对比。

原先，谷仓地面一层是牛棚，上层则用于晾晒和存储干草，放置农场设备。

原有的坡道可以通往养牛区上方的木质露台，现在被保留下来作为通往走廊公寓的主入口。原先临近入口的外部存储区被重新设计为可以俯瞰阿尔卑斯山的观景门廊。所有外部木质包裹层和混凝土屋顶石板都被保留下来，唯一的嵌入结构是前门廊的室内窗户和洞口后的木结构上的孔洞。

沿着房屋主要体量设置的、位于每一个现有的木质结构之间的空间被分别设计成了起居空间、餐厅和抬高的卧室。而像衣橱、浴室、桑

拿间、壁炉和厨房这样的辅助空间则位于服务区内，该服务区位于竖直的厚木板墙后侧。客房位于露台正上方，呈开放性设计，像走廊一样，面向主起居空间。

内部所使用的木材全部来自当地经过深度打磨的云杉。

Alpine Barn Apartment

Traditional landmarks creating a Slovenian countryside also include different types of farmhouses, hay racks and barns.

Unfortunately many of these no longer serve its purpose, therefore are mostly in poor condition, non-maintained and often simply destroyed and replaced with generic housing. Only by embracing the traditions that have been passed on to by their ancestors ensures that the Slovene vernacular architecture survives not just as a national symbol, but also an integral part of the Slovenian rural lifestyle.

Old Barn Revitalization

The concept converts old barn into a loft apartment by leaving the original exterior appearance intact in contrast with creating a new internal wooden shell.

Originally, the ground floor served as stable for stock and upper level for drying and storing hay and farm equipment. The existing ramp, which leads to the wooden deck above cattle area is kept and serves as the main entrance to the new gallery apartment. The former external storage area next to the entrance is converted into the porch overlooking the Alps. All external wooden cladding and concrete roof slates are maintained, and the only intervention is perforations into wooden parts behind internal windows and opening of the front porch.

Along the main volume between each existing wooden structure line are positioned living area, dining and raised bedroom. The auxiliary spaces like wardrobe, bathrooms, sauna, fireplace and kitchen are packed inside the service box displayed on the side behind the wall created by vertical planks. Guest bedroom is created above the terrace, and opened as a gallery towards the main living space.

All internal shell is made of deep-brushed local spruce.

项目名称：Alpine Barn Apartment
地点：Bohinj, Slovenia
建筑师：OFIS Arhitekti
项目团队：Rok Oman, Spela Videcnik, Andrej Gregoric, Janez Martincic, Michele Albonetti, Maria Della Mea, Tomaž Cirkvencic, Pawel Nikkiel, Gözde Okyay, Roberta Costa, Maria Rosaria Ritonnaro, Ralea Toma Ioan Catalin, Grega Valencic, Vlad Popa, Tanja Veselic, Jade Manbodh
结构工程师：Projecta / 机械工程师：MM-term
电气工程师：ES / 照明工程师：Arcadia lightwear
总承包商：Permiz
用途：tourist apartment
用地面积：240m² / 总建筑面积：120m² / 有效楼层面积：140m²
设计时间：2014 / 施工时间：2014 / 竣工时间：2015
摄影师：©Tomaž Gregorič (courtesy of the architect)

体量 volume

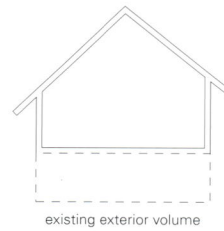

existing exterior volume

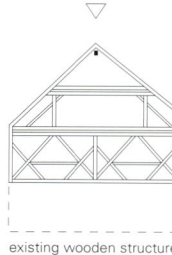

existing wooden structure

interior wooden cladding

living level

bedroom level

service box

1	2	3	4	5	6

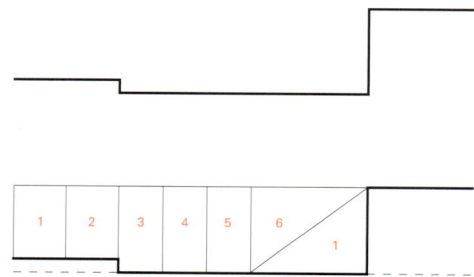

剖面层 section level

1.衣橱　　3.桑拿间　　5.壁炉
2.浴室　　4.厨房　　　6.楼梯间

1. wardrobe　3. sauna　　5. fireplace
2. bathroom　4. kitchen　6. staircase

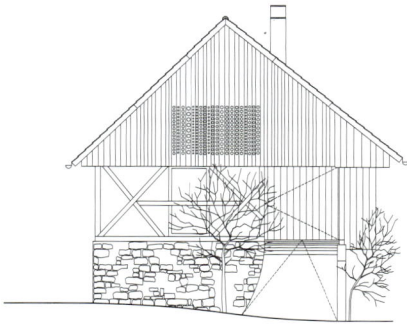

南立面
south elevation

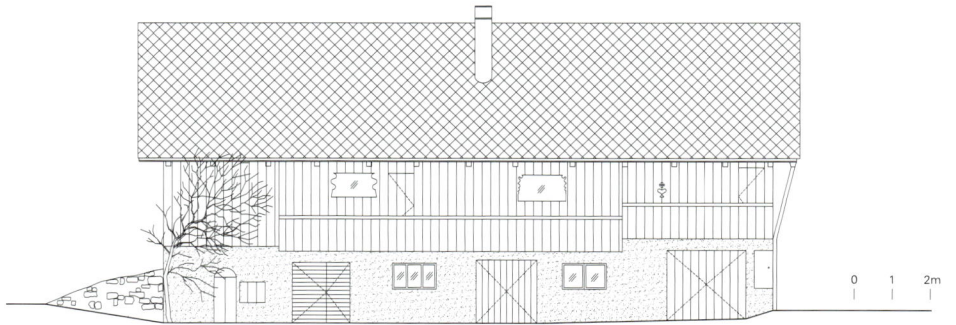

东立面
east elevation

0 1 2m

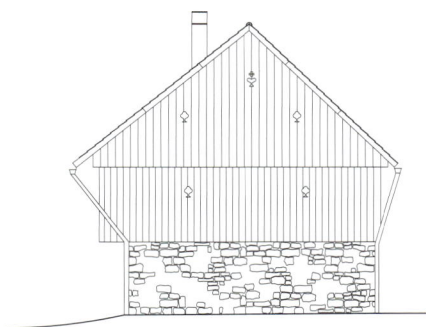

北立面
north elevation

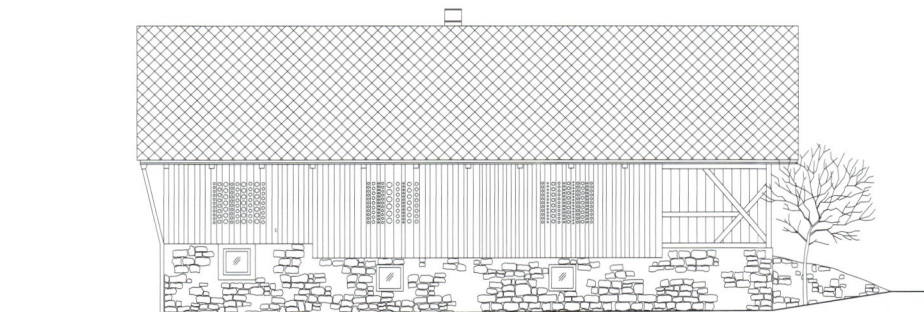

西立面
west elevation

层次理念轴测图
layer concept axo

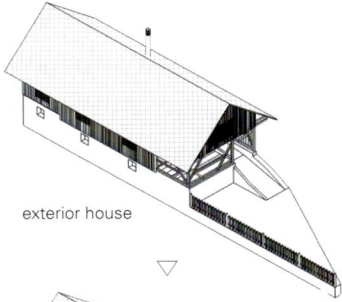

exterior house

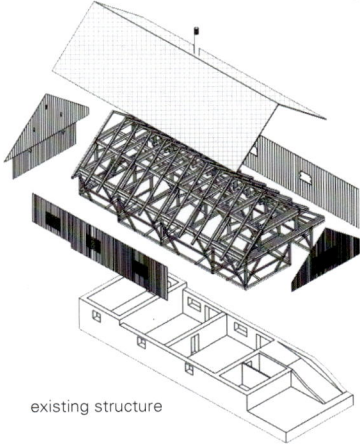

▽

existing structure

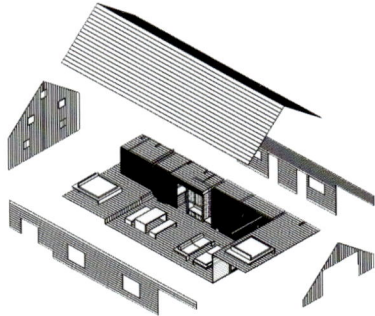

▽

internal wooden shell

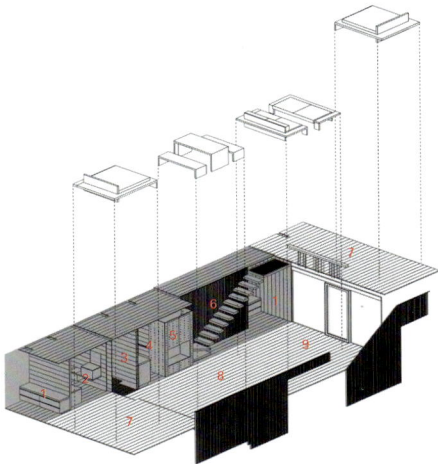

家具 furniture

1.衣橱	6.楼梯间	1. wardrobe	6. staircase
2.浴室	7.卧室	2. bathroom	7. bedroom
3.桑拿间	8.餐厅	3. sauna	8. dining room
4.厨房	9.起居室	4. kitchen	9. living room
5.壁炉		5. fireplace	

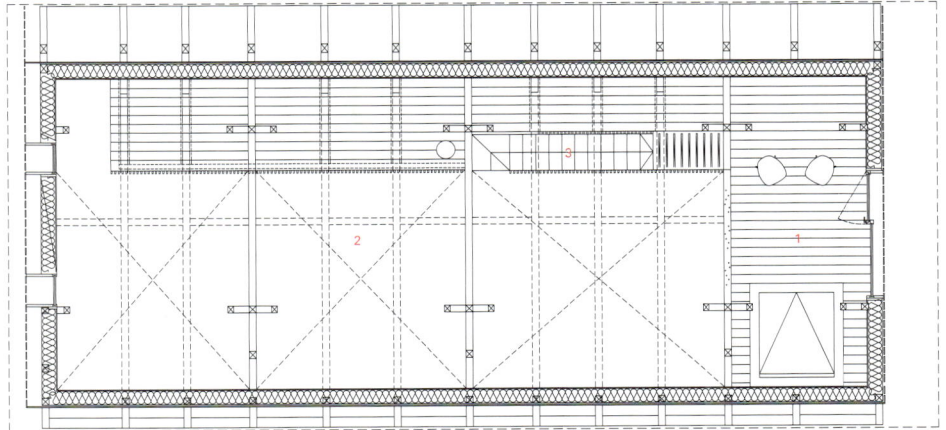

1.卧室 2.双高空间 3.楼梯间
1. bedroom 2. double height space 3. staircase
走廊层 gallery floor

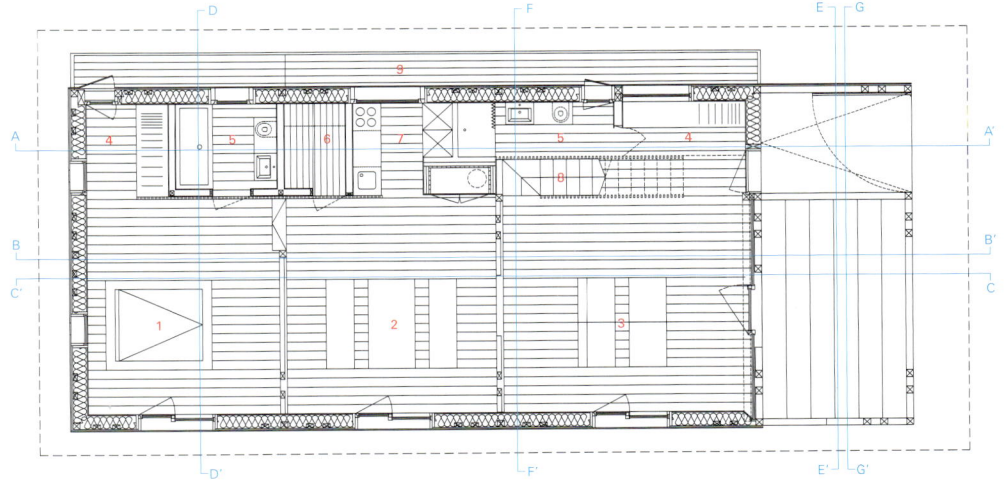

1.卧室 2.餐厅 3.起居室 4.衣橱 5.浴室 6.桑拿间 7.厨房 8.楼梯间 9.露台
1. bedroom 2. dining room 3. living room 4. wardrobe 5. bathroom 6. sauna 7. kitchen 8. staircase 9. terrace
二层 first floor

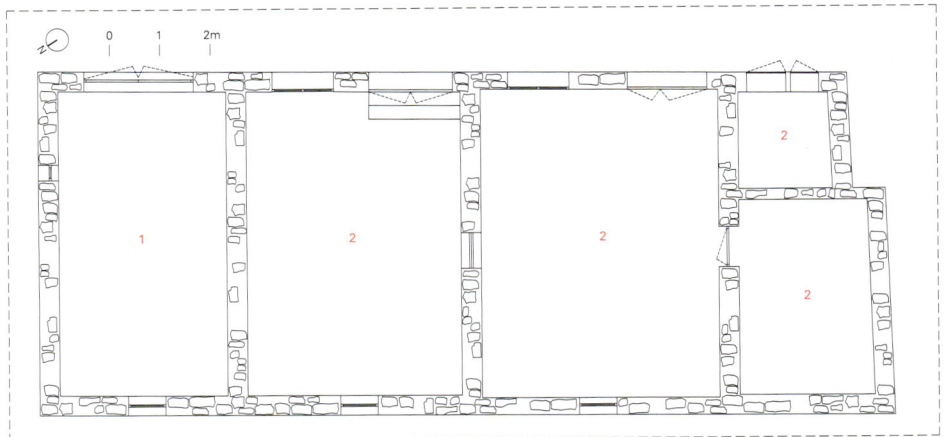

1.车库 2.储藏间
1. garage 2. storage
一层 ground floor

52

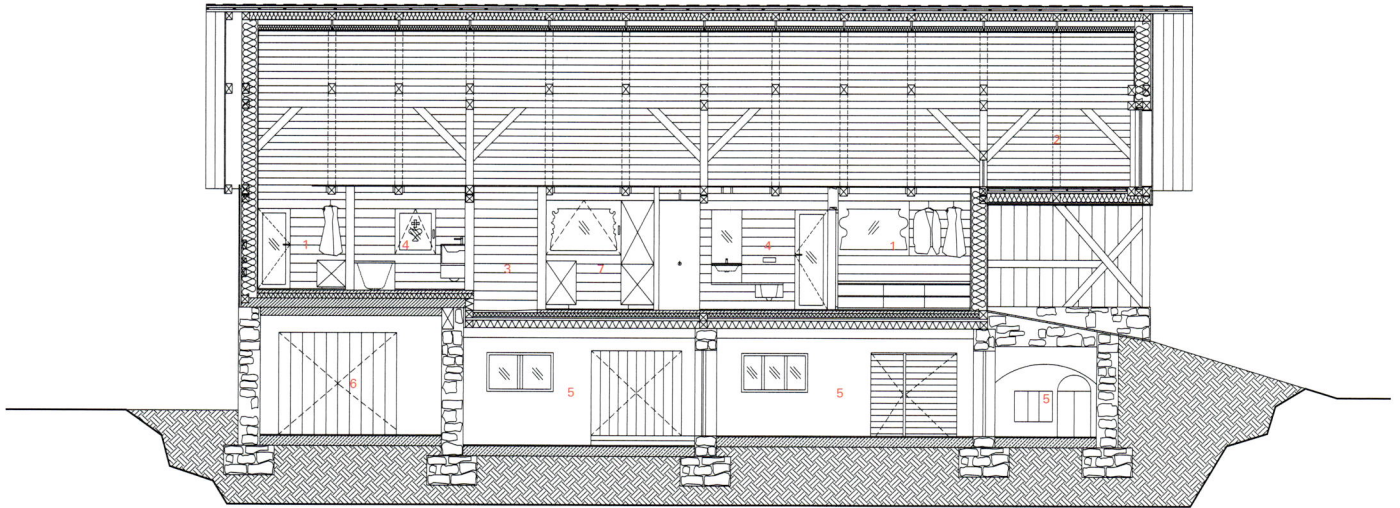

1.衣橱 2.卧室 3.桑拿间 4.浴室 5.储藏间 6.车库 7.厨房
1. wardrobe 2. bedroom 3. sauna 4. bathroom 5. storage 6. garage 7. kitchen
A-A' 剖面图 section A-A'

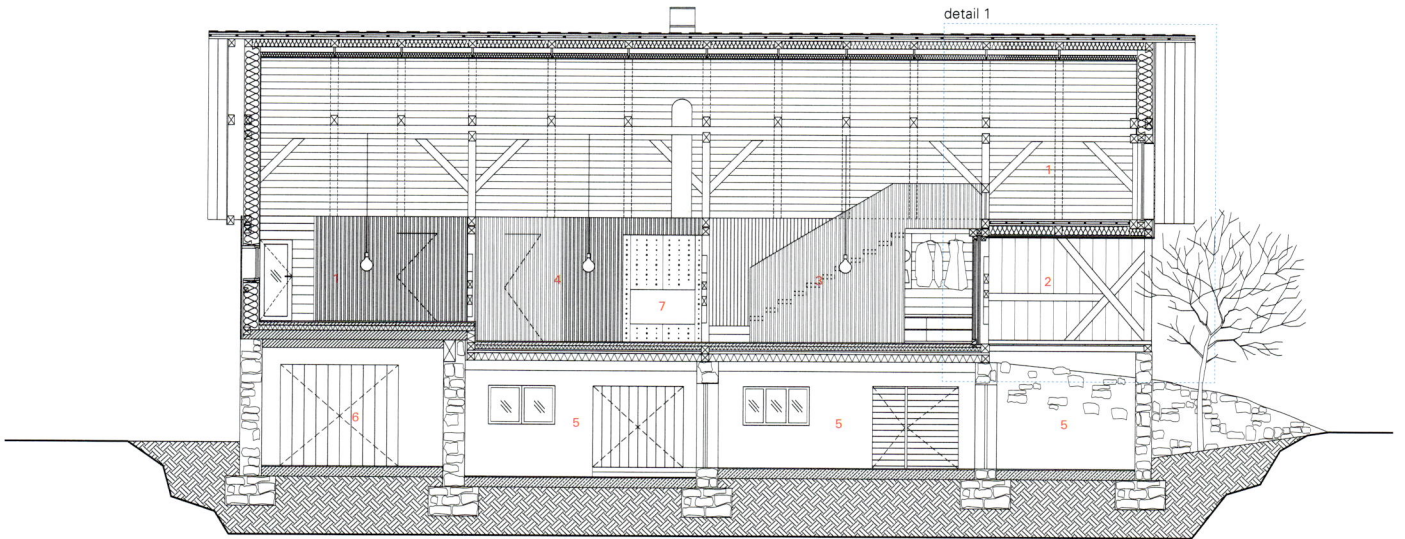

detail 1

1.卧室 2.露台 3.起居室 4.餐厅 5.储藏间 6.车库 7.壁炉
1. bedroom 2. terrace 3. living room 4. dining room 5. storage 6. garage 7. fireplace
B-B' 剖面图 section B-B'

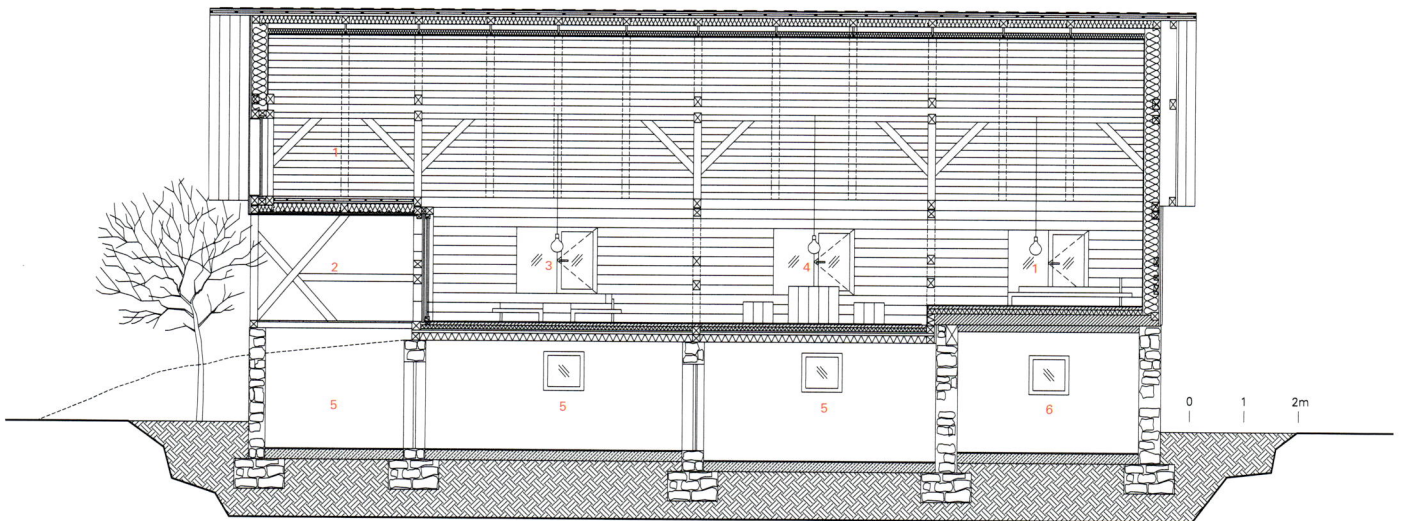

1.卧室 2.露台 3.起居室 4.餐厅 5.储藏间 6.车库
1. bedroom 2. terrace 3. living room 4. dining room 5. storage 6. garage
C-C' 剖面图 section C-C'

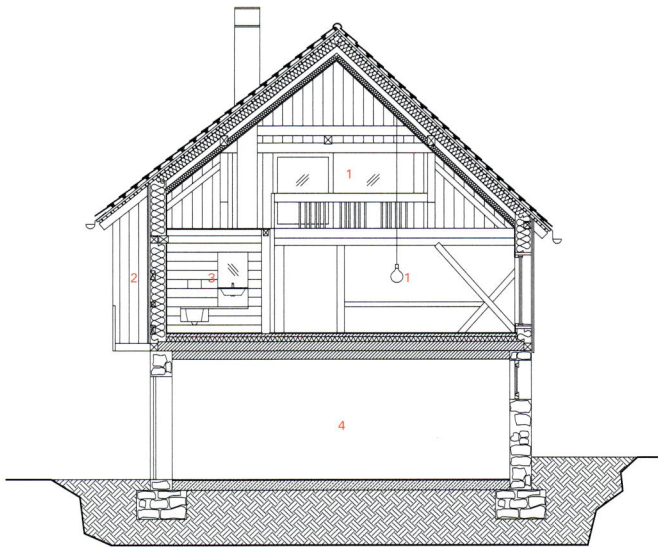

1.卧室 2.露台 3.浴室 4.车库
1. bedroom 2. terrace 3. bathroom 4. garage
D-D' 剖面图 section D-D'

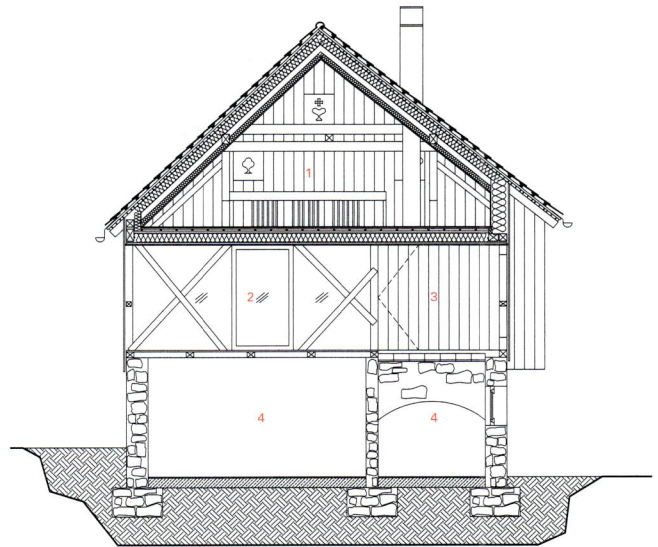

1.卧室 2.露台 3.入口 4.储藏间
1. bedroom 2. terrace 3. entrance 4. storage
E-E' 剖面图 section E-E'

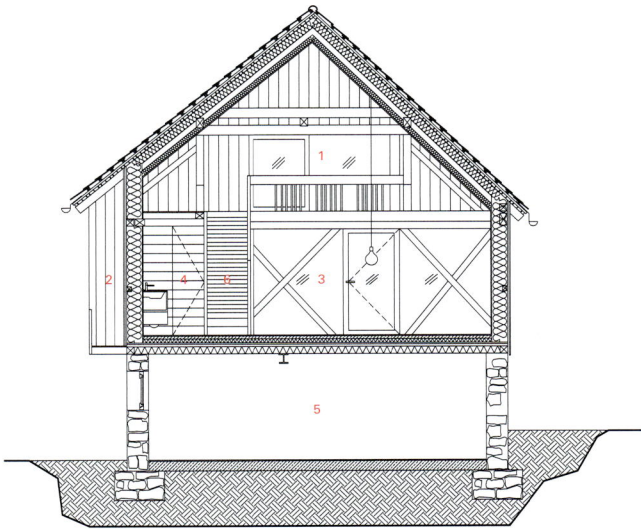

1.卧室 2.露台 3.起居室 4.浴室 5.储藏间 6.楼梯间
1. bedroom 2. terrace 3. living room 4. bathroom 5. storage 6. staircase
F-F' 剖面图 section F-F'

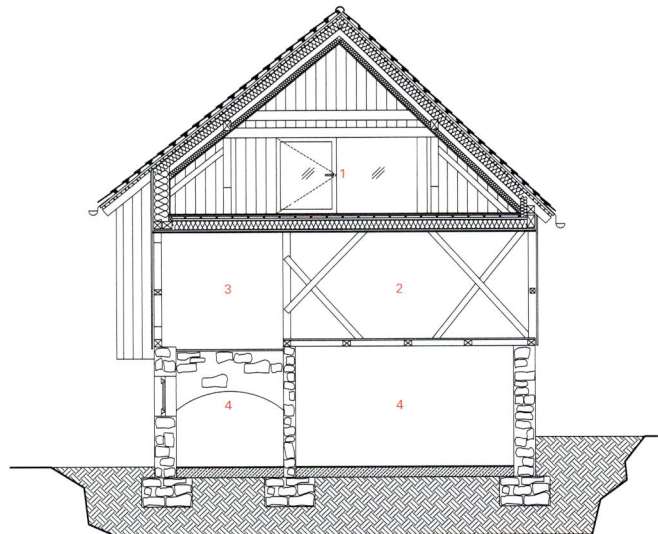

1.浴室 2.露台 3.入口 4.储藏间
1. bedroom 2. terrace 3. entrance 4. storage
G-G' 剖面图 section G-G'

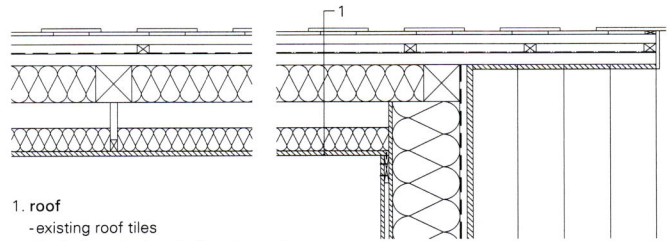

1. **roof**
   - existing roof tiles
   - 2 x timber structure 4x5cm (air gap)
   - hydro insulation foil
   - wooden panels 5x6cm
   - thermal insulation 16cm, 10cm
   - interior wooden cladding 1.5cm
2. **facade**
   - exterior existing wooden cladding
   - thermal insulation 25cm
   - wooden panel 1.5cm
   - wooden sub structure 2cm
   - interior wooden cladding 1.5cm
3. **floor structure (gallery)**
   - deep brushed local spruce boards 2cm
   - oriented strand board 2.5cm
   - wooden sub structure 3x4.5cm
   - existing wooden panels 4cm
   - wooden sub structure for ceiling 3x4.5cm
   - facade foil / exterior wooden ceiling 2.5cm

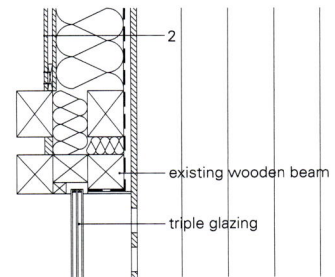

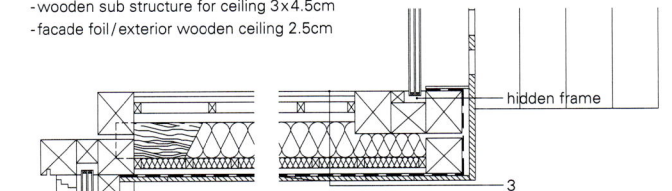

existing wooden beam
triple glazing

hidden frame

triple glazing
triple glazed sliding door

4. **floor structure (first floor)**
   - deep brushed local spruce boards 2cm
   - thermal, sound insulation 2cm
   - thermal insulation 6cm
   - wooden panels 3cm
   - existing wooden panels 5cm
   - thermal insulation 15cm
   - existing wooden panels 3cm
   - existing metal beam

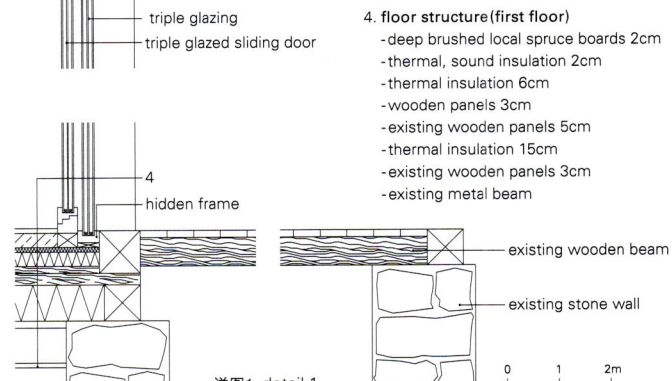

hidden frame

existing wooden beam

existing stone wall

0   1   2m

详图1 detail 1

弗罗林斯公寓

Baumhauer

位于弗罗林斯这一瑞士小村庄中的古典的恩加丁农舍被Philipp Baumhauer建筑事务所在尊重原有建筑的基础上进行了重新整修和改造，它由两部分组成，面朝村庄中心的居住部分和远离公路的农场部分。

该建筑被完全重新翻新，增加了额外的居住空间。首先，建筑师把房子后面原来用作谷仓的部分中的地板全部拆除，然后将一个三层的独立木质结构嵌入到原有的石头框架内，人们只能从外立面上的大大的木质体量看出内外差异。尽管千变万化，但该设计使古老的巨大谷仓体量仍清晰可辨。二层高的起居室的屋顶向上延伸至嵌入结构的屋顶，其内表面全部为落叶松木材饰面。

旧谷仓建筑内又额外新建了四个房间。

整个房屋沿着纵向轴将视野全部开放，一扇巨大的玻璃拉门通向阳台，使人们饱览恩加丁山脉的优美风光。设计在厚重的外墙中嵌入当地典型的深层视窗造型，并进行了夸张处理，表明建筑里面进行的改变。仅有的几个巨大的窗户洞口边缘经过清晰的细化处理，与原有建筑形成鲜明对照。建筑外立面参考当地的古老谷仓，选择使用粗锯的、没有经过处理的落叶松木板饰面。木材不断风干老化的过程有助于房屋与周围原有建筑物融为一体。

Florins Residence

This classic Engadine farmhouse in the Swiss hamlet of Florins was renovated and converted by Philipp Baumhauer Architekten with respect for the existing. It consisted of a residential section facing the village center and an agricultural section away from the road.

The building was completely refurbished and additional living space was added. By first removing the floor of the back barn section of the house, a three-level, structurally independent wood structure, was then inserted within the existing stone frame, distinguishing itself on the exterior facade as a large wooden volume. The design allows the historical large volume of the barn to remain legible despite the changes. The two-story living room extends up to the roof of the inserted structure, whose inner surfaces are clad with larch wood. Four

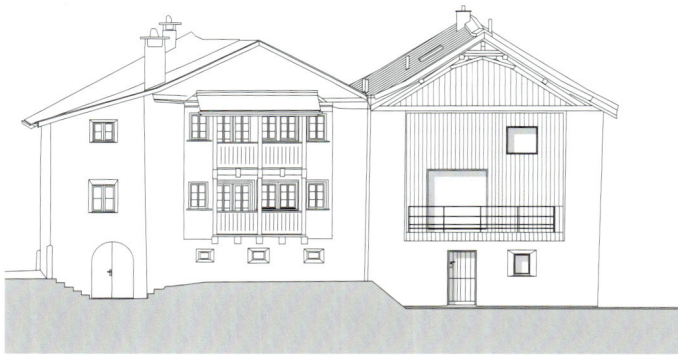

南立面 south elevation

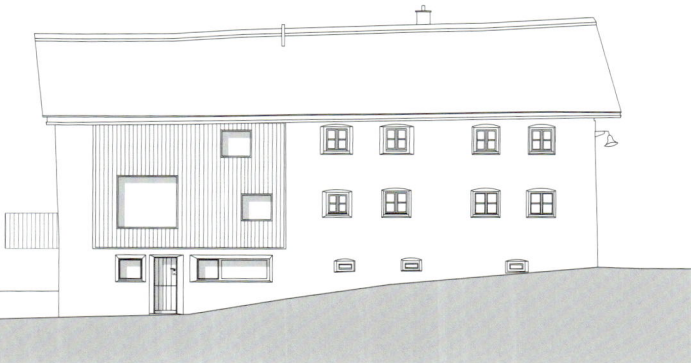

东立面 east elevation

0 1   3m

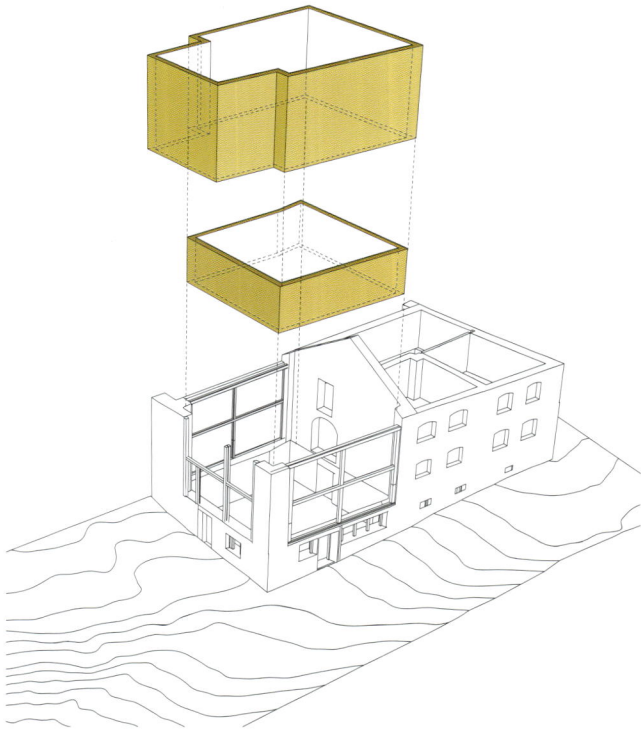

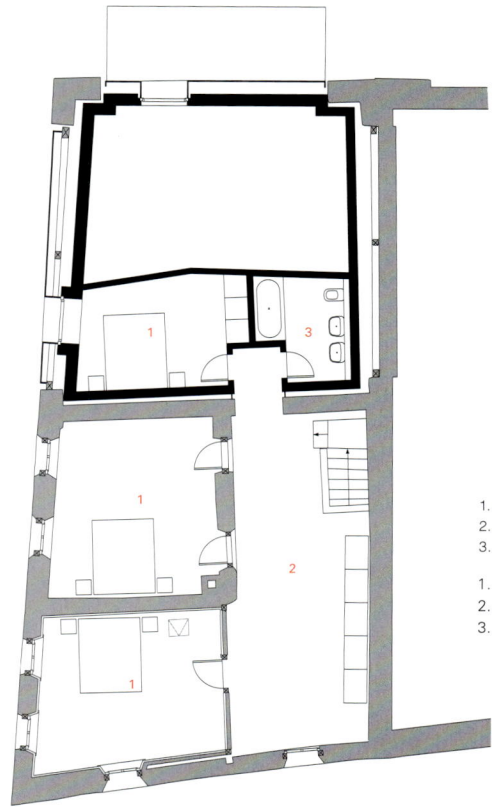

三层　third floor

1.卧室
2.上层走廊
3.浴室

1. bedroom
2. upper corridor
3. bathroom

一层　first floor

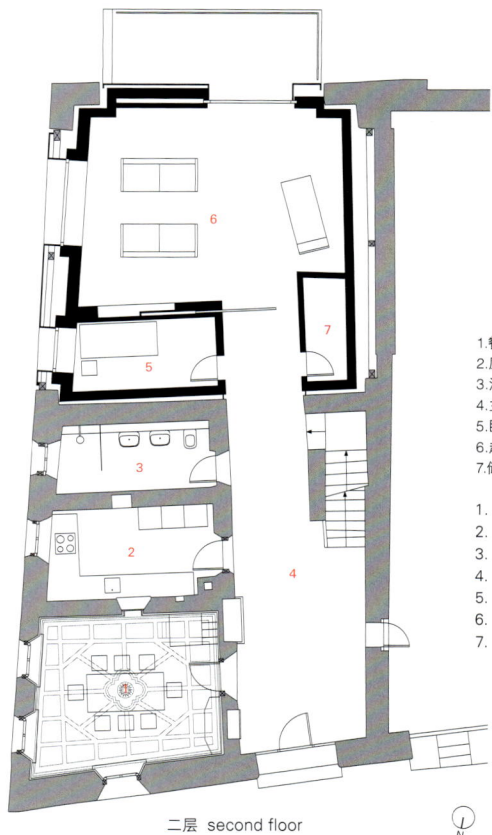

1.酒窖
2.地窖
3.底层走廊
4.书房
5.前厅
6.浴室
7.杂物房
8.洗衣房

1. wine cellar
2. cellar
3. lower corridor
4. study room
5. mud room
6. bathroom
7. utility room
8. laundry room

二层　second floor

1.餐厅
2.厨房
3.浴室
4.主走廊
5.卧室
6.起居室
7.储藏间

1. dining room
2. kitchen
3. bathroom
4. main corridor
5. bedroom
6. living room
7. storage

N　0　1　2m

additional rooms were also built within the old barn. Views are open through the entire house along its longitudinal axis, and a large sliding glass door to the balcony extends the view to the mountains of the Engadine. The typical local motif of deep window revealed in thick exterior walls has been adopted and exaggerated, signifying the alterations inside. With their sharp-edged detailing, these few large window openings stand out in contrast to the existing architecture.

Referencing historical barns in the vicinity, rough-sawn, untreated larch boards were chosen for the facade. The ongoing aging process of the wood helps the project to fit within the built fabric of its surroundings.

项目名称：Florins Residence
地点：Tarasp, Scuol, Switzerland
建筑师：Baumhauer
设计团队：Philipp Baumhauer, Julian Sauer, Tomohiro Yanagisawa, Kevin Driscoll
总建筑面积：524m²
竣工时间：2015
摄影师：©Ralph Feiner

0   0.2   0.5m

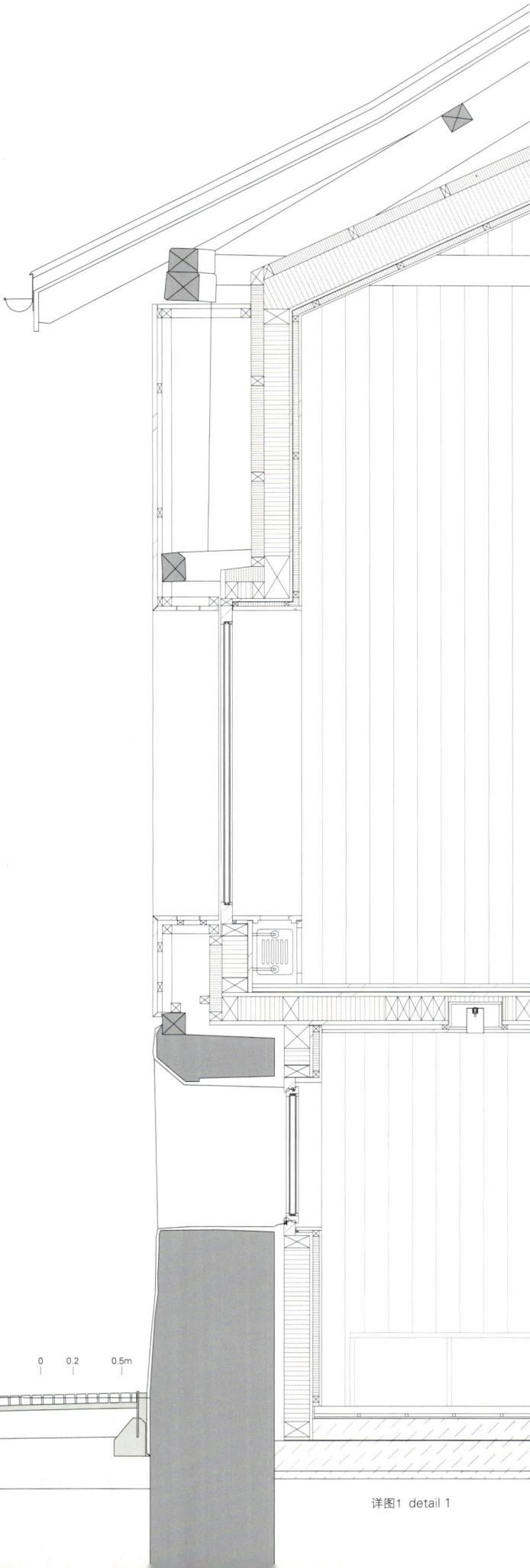

详图1 detail 1

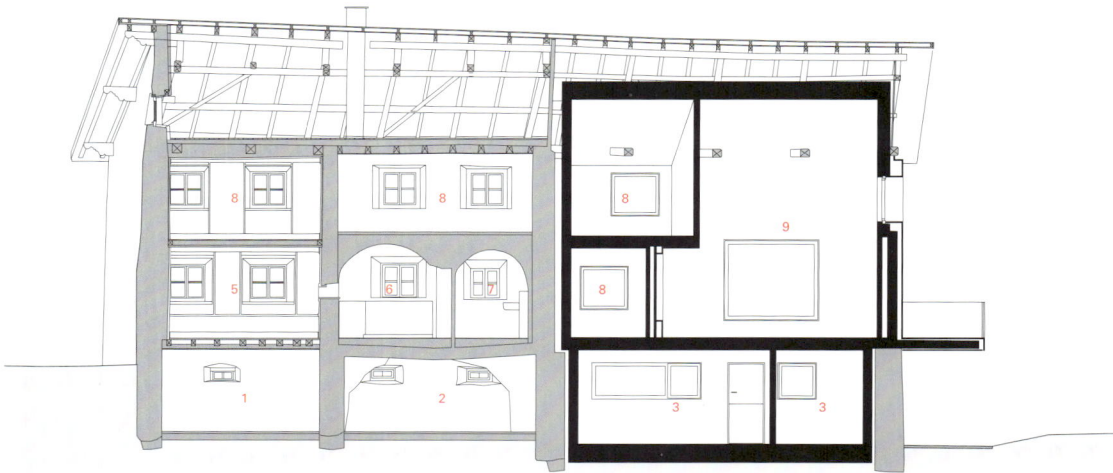

A-A' 剖面图 section A-A'

1. 酒窖
2. 地窖
3. 书房
4. 前厅
5. 餐厅
6. 厨房
7. 浴室
8. 卧室
9. 起居室
10. 储藏间
11. 洗衣房

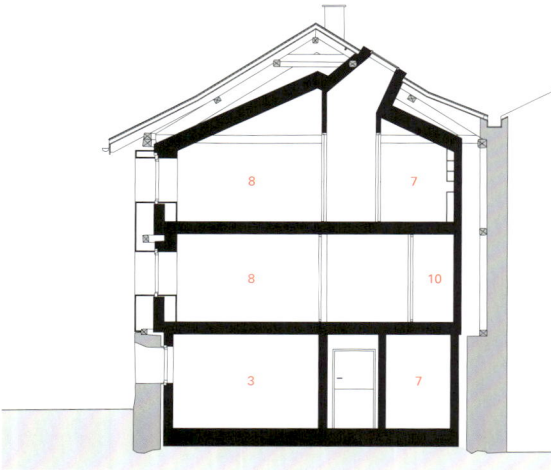

B-B' 剖面图 section B-B'

detail 1

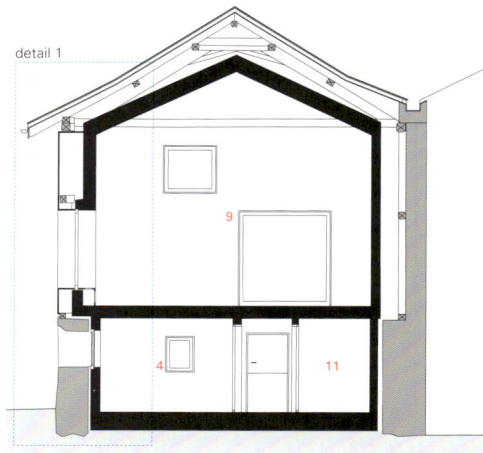

C-C' 剖面图 section C-C'

1. wine cellar
2. cellar
3. study room
4. mud room
5. dining room
6. kitchen
7. bathroom
8. bedroom
9. living room
10. storage
11. laundry room

0   1   2m

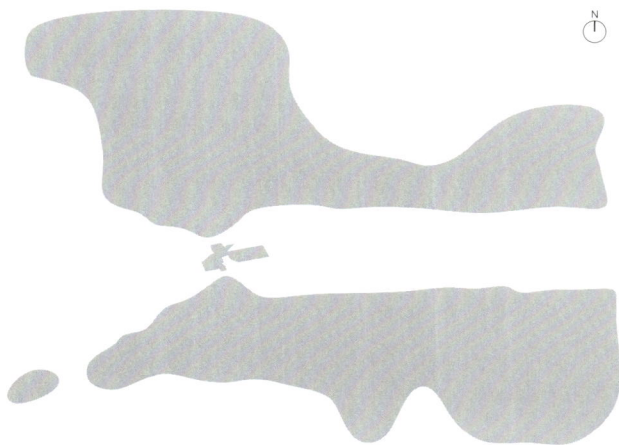

23.2大宅

Omer Arbel

23.2大宅是为一户人家建的住宅, 位于农村, 占地面积很大, 东西向的缓坡和两处森林定义了截然不同的"户外房间"。该建筑位于这两种环境的交界点, 因此既定义了两处的完全不同特色, 也是其间的一个集中过渡点。

本设计的出发点是把房屋作为回收利用的、具有百年历史的花旗松横梁的储藏所而设计。所有这些横梁的每一根都铣自于一棵单独的树, 因此长度和横截面尺寸各不相同, 有的长达20m, 有的横截面直径有1m。该项目把这些横梁看作是当地具有丰富社会和生态历史的考古文物来对待。因此, 不会对其进行铣削、切割或者抛光打磨。因为横梁的长度和大小不同, 所以建筑师设计了一个自由三角形几何体结构来适应横梁的尺寸, 由此在室内外空间营造出显性和隐性的关系。

为了最大化地模糊室内外的分界线, 该项目采用折叠门系统, 可以让房屋前后两侧的外墙全部折叠消失, 每个房间的角落也不复存在。

23.2 House

23.2 is a house for a family, built on a large rural acreage with a gentle slope from east to west and two masses of forest defining distinct "outdoor rooms". The house is situated at the point of transition between these two environments, and as such acts at once to define the two as distinct, and also to offer a focused transition between them.

The project began, as a point of departure, with a depository

项目名称：23.2 House
地点：Canada, Vancouver
建筑师：Omer Arbel
用地面积：161,874m²
总建筑面积：465m²
竣工时间：2010
摄影师：©Nic Lehoux (courtesy of the architect)

of reclaimed, century old Douglas Fir beams. The beams were each milled of a single tree, and consequently were of different lengths and cross sectional dimensions, some as long as 20m, some as deep as 1m. The project treats the beams as archaeological artifacts of the social and ecological history

of the region. As such, they were not milled, cut, or finished. Because the beams were of different lengths and sizes, a free triangular geometry was developed to accommodate the variety in dimension and to create implicit and explicit relationships between indoor and outdoor space.

In order to maximize ambiguity between interior and exterior space, the project employed a strategy of removing one significant corner of each room, with an accordion door system on both sides, allowing the entire facade to retract.

74

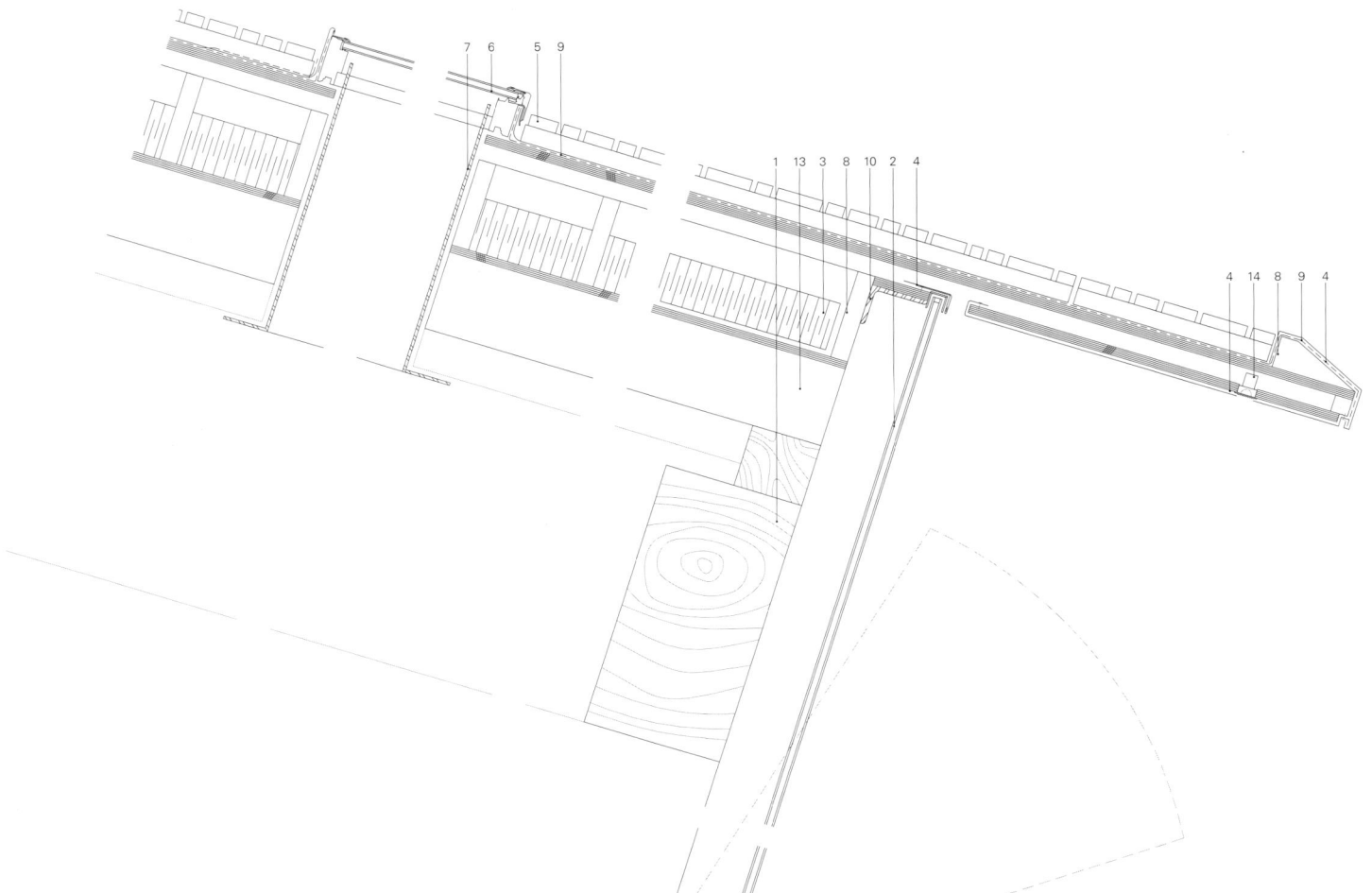

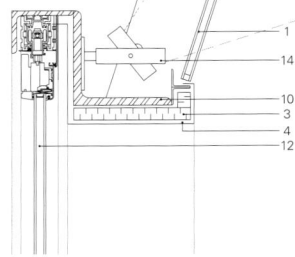

1. reclaimed wood beam
2. double glazed unit
3. insulation
4. steel flashing
5. cedar roof cladding
6. skylight
7. powder coated steel frame
8. framing
9. membrane
10. steel profile
11. concrete topping with radiant heat
12. operable accordion door
13. wood ceiling
14. lighting
15. concrete slab
16. concrete foundation
17. exterior concrete paver

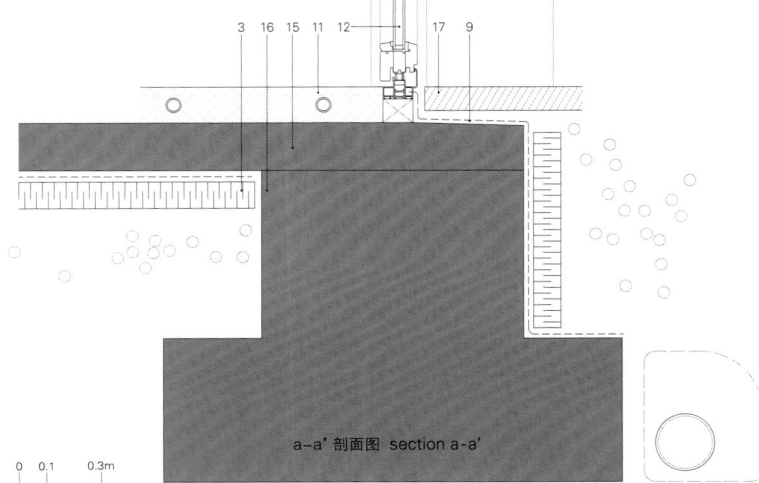

a-a' 剖面图　section a-a'

0　0.1　0.3m

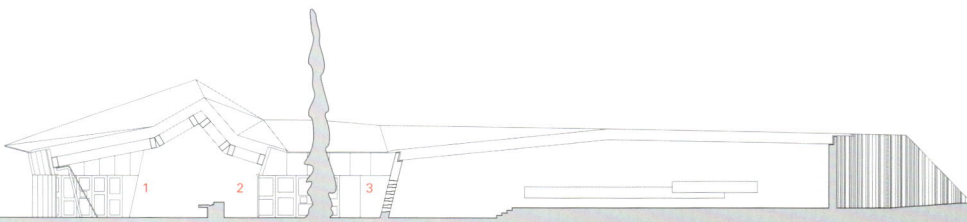

1.大房间 2.厨房 3.庭院 1. great room 2. kitchen 3. courtyard
A-A' 剖面图 section A-A'

0 2 5m

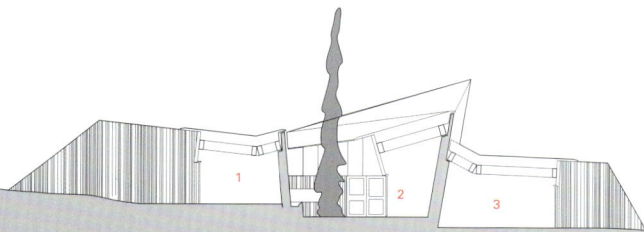

1.入口 2.庭院 3.主浴室 1. entry 2. courtyard 3. master bath
B-B' 剖面图 section B-B'

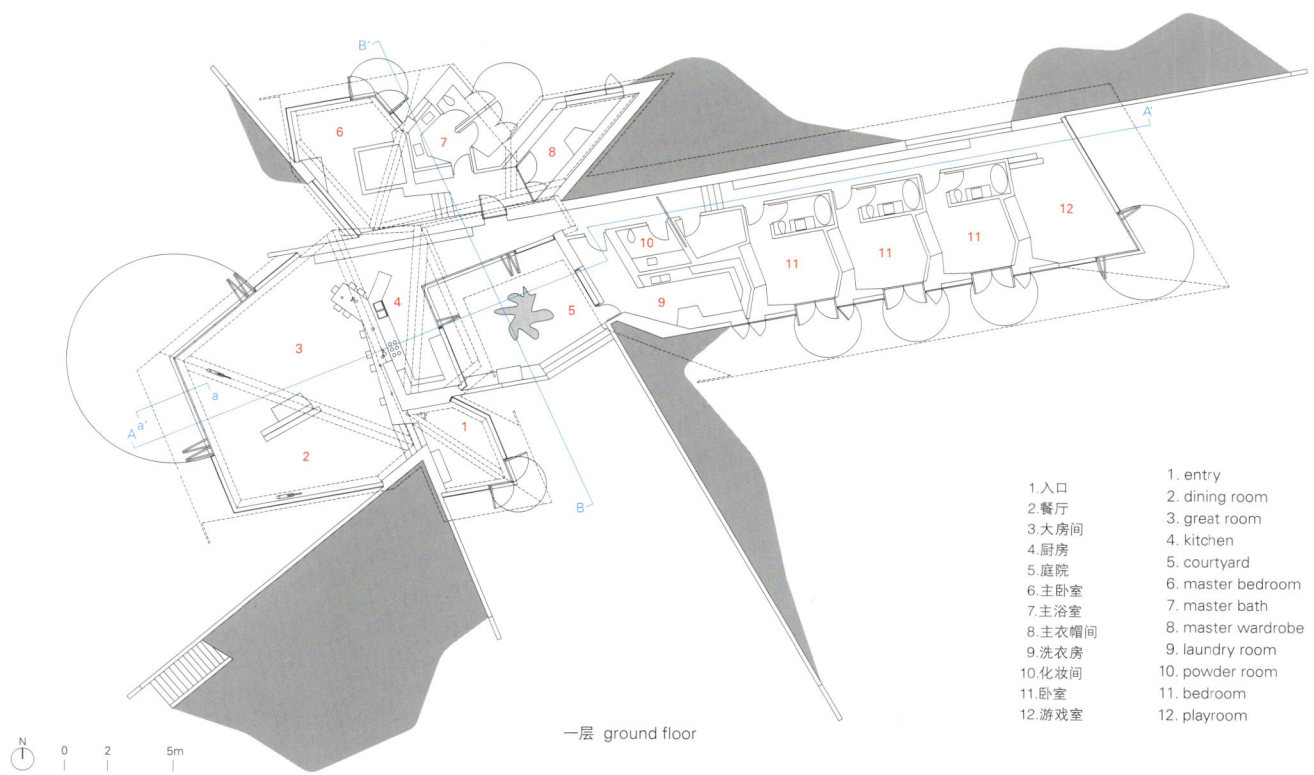

一层 ground floor

1.入口
2.餐厅
3.大房间
4.厨房
5.庭院
6.主卧室
7.主浴室
8.主衣帽间
9.洗衣房
10.化妆间
11.卧室
12.游戏室

1. entry
2. dining room
3. great room
4. kitchen
5. courtyard
6. master bedroom
7. master bath
8. master wardrobe
9. laundry room
10. powder room
11. bedroom
12. playroom

N 0 2 5m

鸠谷屋
Schemata Architects

　　我们的客户从他父亲手里继承了这座双层木屋，委托我们加以改造。客户在上小学时住在这座房子里，后来该房屋处于长期出租的状态。这座房屋始建于社会大发展、人口极速增长时期，当时谁都没想到会出现由人口下降导致的社会问题，如"房屋无人居住"。原来房屋的设计是在有限的空间内尽可能多地提供私人专用房间。此外，建筑师的设计完全是在客户的父亲的指示下完成的，所以整座住宅都强烈地反映了他父亲的意愿。每间屋子的风格都不一样，细节上有很大差别，窗户的形状也各不相同。这种对房子的强烈意愿或者说是"爱"似乎有点沉重。为了缓和这种沉重感，我们拆除了一些空间，整理出某些具有共同特征的组成部分，营造出一种整体感——一种新形式的"爱"——这也是我们想在房间里呈现的东西。

　　由于预算有限，我们无法对原有窗户进行改动。但是，我们给每个独特的窗户安装了内墙，使其看起来像一件艺术作品。另外，我们拆除了二层的部分楼板，设计了一处上空空间。这处"上空"将所有原本关上门就与其他房间隔离开来的房间连为一体。上空空间在一层，夹层和二层之间起到了很好的协调作用，重新配置了整个房屋内的空间关系。该"上空"空间还起到采光井的作用，通过南墙上的洞口将自然光洒满室内。

House in Hatogaya

Our client took over this 2 story wooden house from his father and commissioned us to work on the renovation. The client lived in this house during his elementary school days, and it had been rented for a long time afterwards. It was built during the time of social and population growth when nobody imagined current social issues such as "vacancy" resulted from population decline would arise. The house was designed to provide as many private rooms as possible within the limited space. In addition, the design was executed by an architect under his father's directions, and his father's strong intentions were reflected all over the house. Each room had a different

taste, with particular details and uniquely shaped windows. Accumulation of such strong intentions or "love" for the house felt a bit too heavy. In order to alleviate such heaviness, we removed some of spacial components, sorted out some components with common characters, and created a sense of integrity – which is a new form of "love" – we intend to present in this house.

We could not modify the existing windows, due to the limited budget. Instead we framed each unique window with an interior wall, reconfiguring it like a piece of artwork. And we punched out a floor of a second floor room to create a "void" connecting all rooms which had been separated behind closed doors. The "void" mediates between the first floor, the mezzanine, and the second floor, re-configuring spatial relationships throughout the house. The "void" also functions as a light well distributing the natural light through openings on the south wall. Schemata Architects

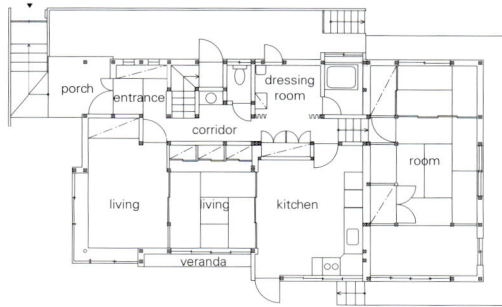

一层_改造前 first floor_before

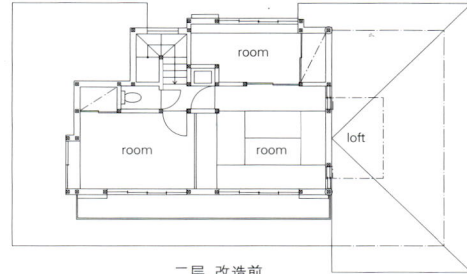

二层_改造前
second floor_before

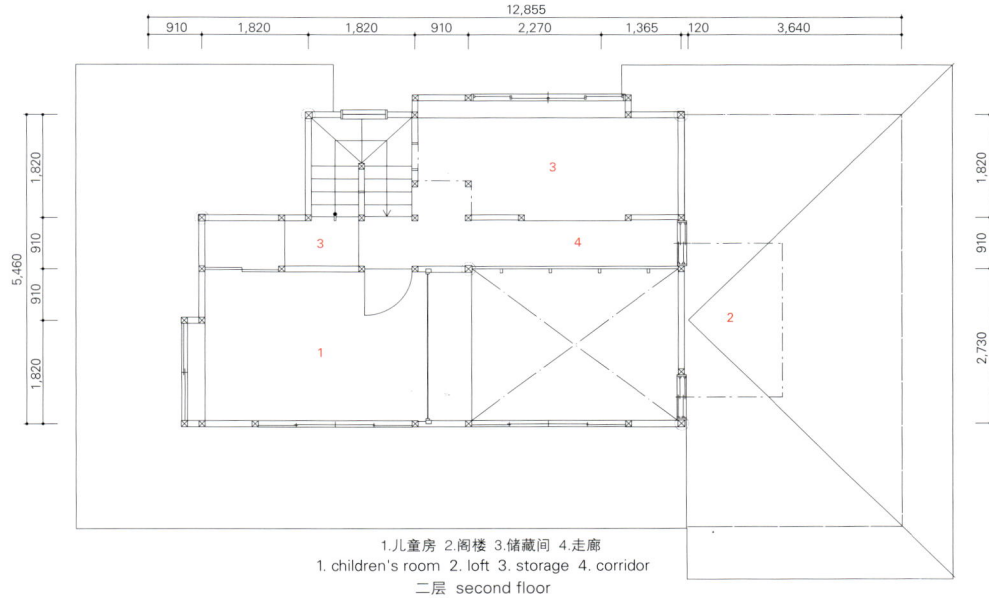

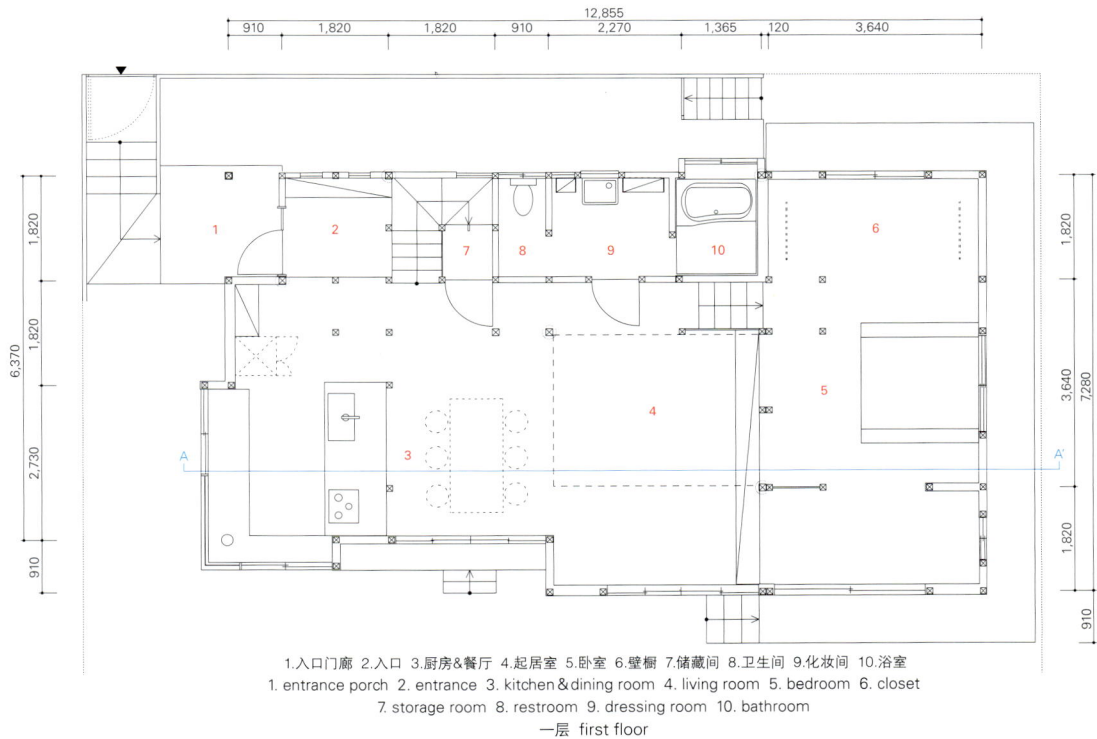

1.儿童房 2.阁楼 3.储藏间 4.走廊
1. children's room 2. loft 3. storage 4. corridor
二层 second floor

1.入口门廊 2.入口 3.厨房&餐厅 4.起居室 5.卧室 6.壁橱 7.储藏间 8.卫生间 9.化妆间 10.浴室
1. entrance porch 2. entrance 3. kitchen&dining room 4. living room 5. bedroom 6. closet
7. storage room 8. restroom 9. dressing room 10. bathroom
一层 first floor

项目名称：House in Hatogaya / 地点：Hatogaya, Kawaguchi City, Saitama Prefecture, Japan
建筑师：Jo Nagasaka / 项目团队：Reina Sakaguchi / 施工单位：TANK / 用途：private house
用地面积：300.45m² / 有效楼层面积：87.921m² / 总建筑面积：116.16m² / 总建筑规模：one story below ground,
two stories above ground / 结构：wood + reinforced concrete / 设计时间：2014.10~2015.4 / 施工时间：2015.4~7
竣工时间：2015.8 / 摄影师：©Kenta Hasegawa (courtesy of the architect)

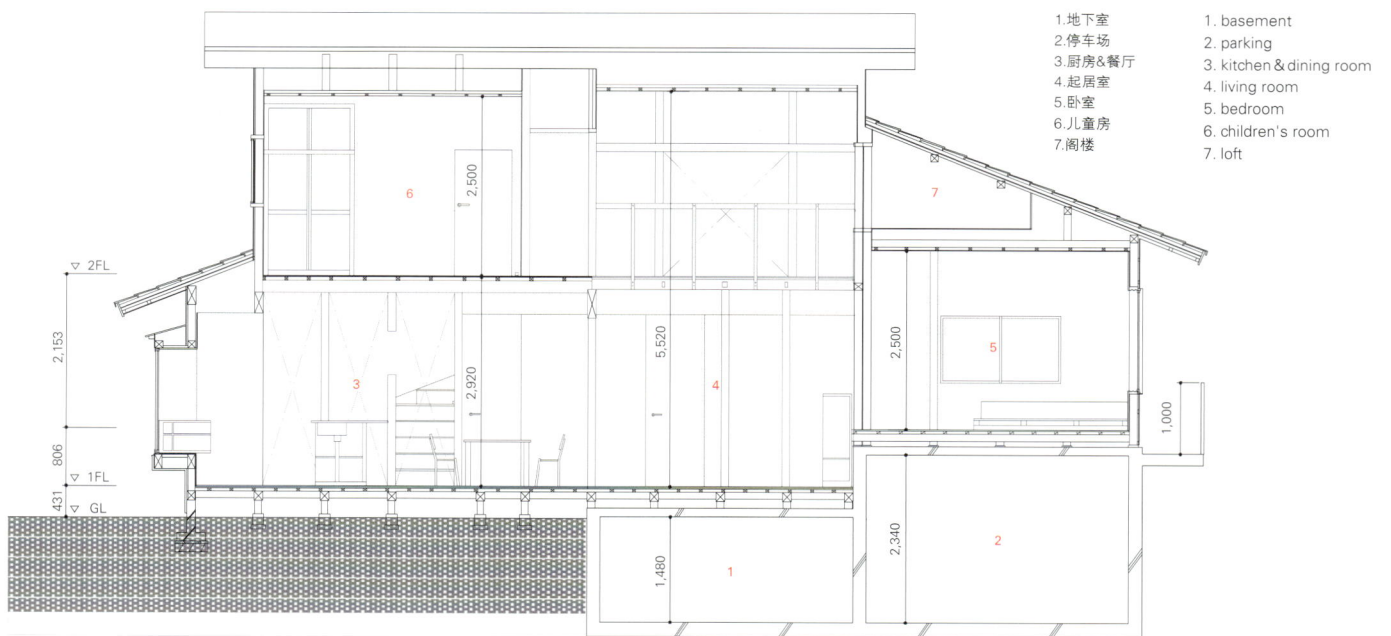

1.地下室
2.停车场
3.厨房&餐厅
4.起居室
5.卧室
6.儿童房
7.阁楼

1. basement
2. parking
3. kitchen & dining room
4. living room
5. bedroom
6. children's room
7. loft

A-A' 剖面图 section A-A'

FA住宅

atelier tho.A

这座房子自从房主搬走后已经空了二十多年, 现如今他们想搬回这座老房子住。一开始, 房主想修建一座和老房子相像的新房子, 因为老房子已经严重败落, 甚至不能遮风挡雨。但因为这所房子是房主去世的父亲自己修建的, 出于怀旧的情怀, 建筑师想尽力保留原有的建筑。所以, 最后建筑师决定采用透明材料——聚碳酸酯来覆盖整座旧建筑, 一来保护旧建筑, 二来可以多出一些绿色空间。此外, 这种材料完全满足这个项目的要求: 屋内有阳光, 房主可以随意种树, 而且这里处于高地, 全年凉爽宜人。

由于使用了聚碳酸酯材料, 因此这座古老的建筑从外面看依然古风犹存。旧建筑内部的功能几乎和以前一样, 但是对其中的一些细节进行了改建, 不仅是为了符合新的要求, 也是为了与现代建筑风格相符。主色调仍然是这座房子最初的颜色。

为了展示旧建筑内部的岩石墙, 建筑师选择在一个角落拆掉其表面的砂浆层。这项传统的岩石墙建筑技术来自房主的爷爷。房子的每块石头都是由房主去世的父亲精心挑选并一丝不苟堆砌的。此外, 这个角落还记录了这座房子具有纪念意义的时期: 第一次建成时还只是他们一对夫妻居住, 而第一次扩建时他们已经有了孩子, 第二次翻修是为了养猪等等。

出于安全考虑, 所有旧瓦片都被拆除了, 因为这些瓦片太旧了, 有时掉下来会很危险。内部几乎所有空间都使用了木质天花板, 所以就不需要用新的瓦片更换。旧瓦片用来铺设人行道, 不仅美观, 还能防止用于浇灌的水弄湿脚或鞋。

旧房子的墙与透明塑料墙之间的空间用来工作、就餐或放松休闲。新房子的所有室外门都是由可循环利用的木材制成的, 这种材质成本低廉, 经久耐用。

From 1985 to 2014:
The house had been built with 6x6 m² for a couple. After a couple of years, the house was extended when they had babies. In 1991, the father died, and the house has been abandoned since then.

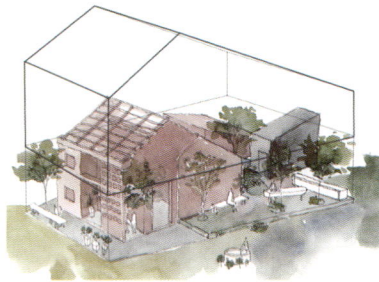

... to 2015
The family wants to return to the old place, to the memories and stay away from crowded. They try to retain the old parts as much as possible but also fit to the modern needs. Some spaces and trees are added to the old house. The result, they have the new farmhouse.

... In future :
They wrap the house by a polycarbonate skin to help the trees grow well and protect the house. The polycarbonate skin can easily adapt the climate in Dalat - the highland of Vietnam – which has the cool temperate over year.

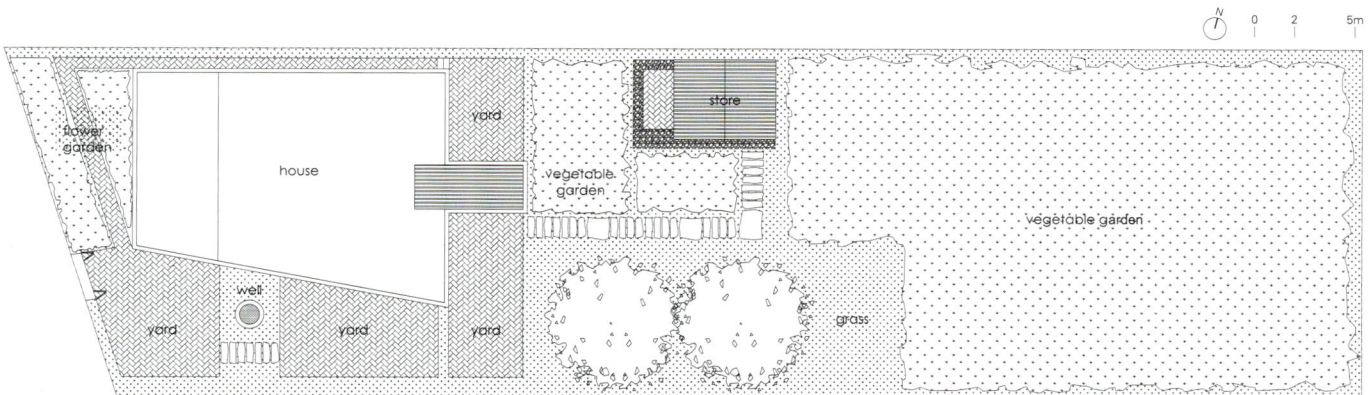

N 0 2 5m

FA House

This house was unused for more than 20 years – after a happening that the owners moved to another place, and now they want to return to the old house. At first, they intend to build a new house looks like the old one, because this place has been seriously degraded and can not protect itself from sunlight or rain. But the architects want to keep the old house as much as they can, because this house was built by the deceased father himself, so it likes a reminiscence. So at last the architects decide to cover all the old building by a transparent material – polycarbonate – to protect the old house and make some new green spaces. Moreover, this material really meets requirements with this case – with the sunlight inside the house, and the owner can plant tree anywhere but it is not too hot because this highland place has cool temperature the whole year.

Due to the use of polycarbonate, the old buildings still can be

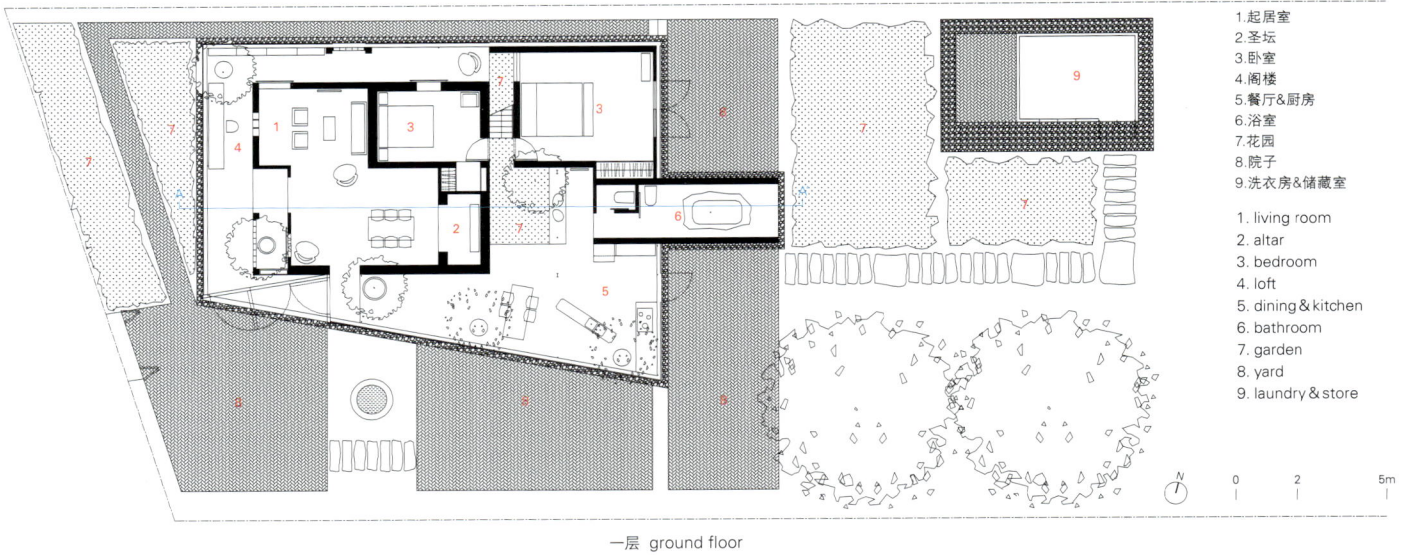

一层 ground floor

1.起居室　1. living room
2.圣坛　2. altar
3.卧室　3. bedroom
4.阁楼　4. loft
5.餐厅&厨房　5. dining & kitchen
6.浴室　6. bathroom
7.花园　7. garden
8.院子　8. yard
9.洗衣房&储藏室　9. laundry & store

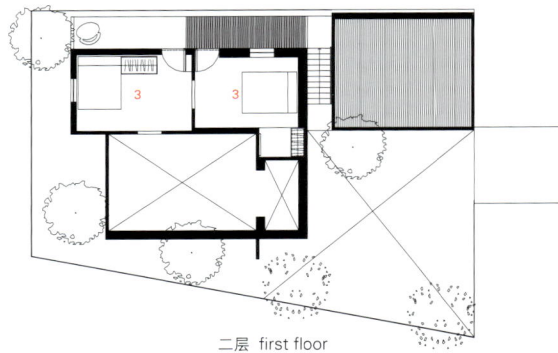

二层 first floor

seen faintly from outside. Inside the old building the functions are almost kept like before but some details are rebuilt in order to suit not only new requires but also contemporary architecture. The dominant color is learned from the first color of this old house.

The architects choose a corner to remove the old mortal in order to show the rock wall inside – this is a traditional technique of building from the owner's grandfather. Each stone was chosen and placed meticulous by the deceased father. Moreover, this corner can also show many memorial periods of this house: the time it was built when they were just a couple, the first time it was expanded when they had

children, and the second time it was renovated to feed some pigs and etc..

All the old tiles had to remove because of safety concerns – they are too old and sometimes can drop down dangerously. Almost spaces inside have wooden ceiling so it's no need to be replaced by new tiles. The old tiles were used to pave walkway, not only for aesthetic purpose but also for preventing feet or shoes from being wet by planting water.

The spaces between two layers – wall of old building and transparent plastic wall – are used for working, dinning and relaxing. All exterior doors of this new house are made from recycled wood – this material with both lower cost and more durability to weather.

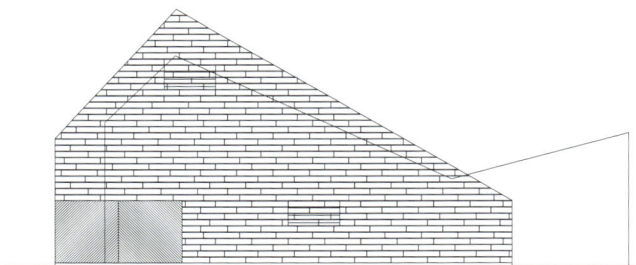

南立面 south elevation

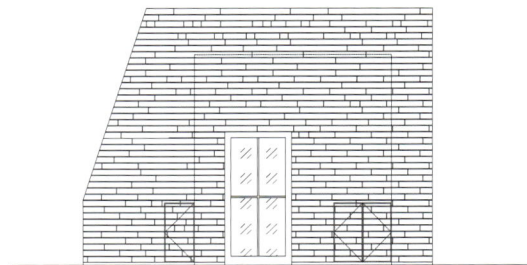

东立面 east elevation

0　2　5m

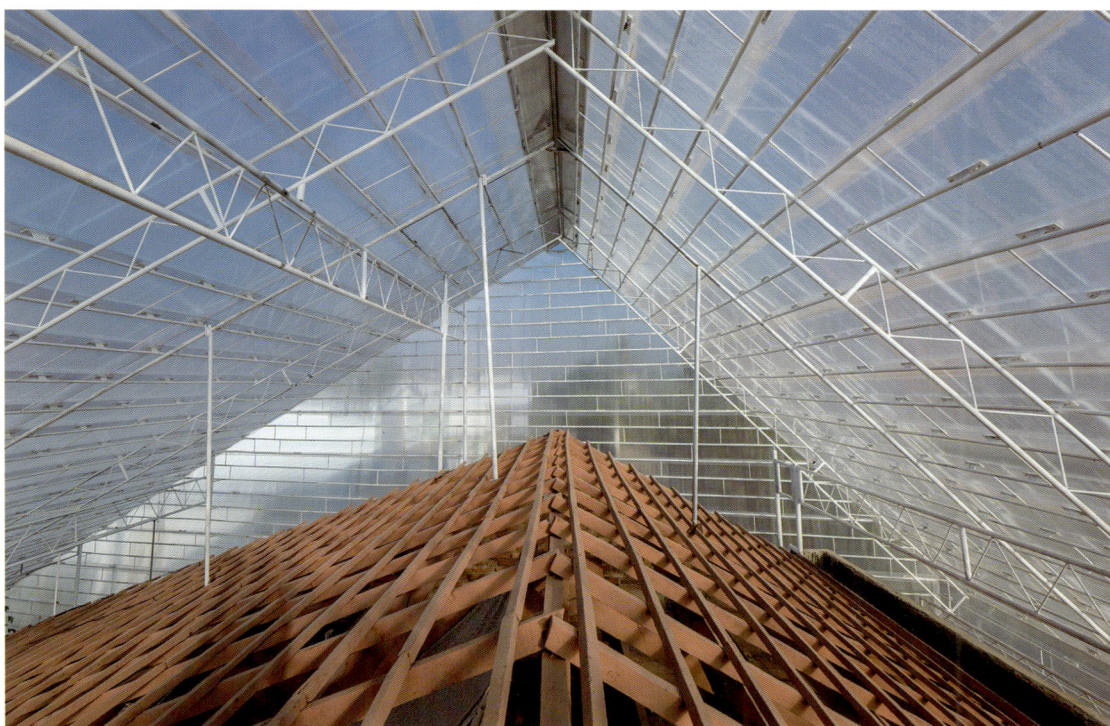

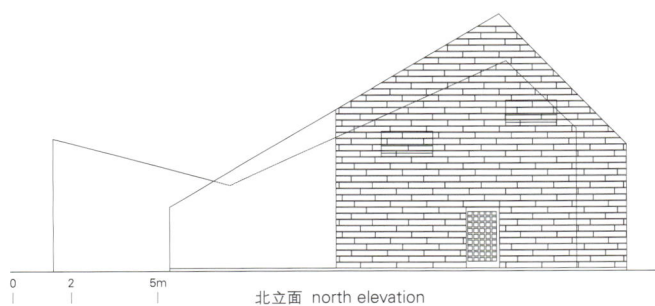

北立面 north elevation

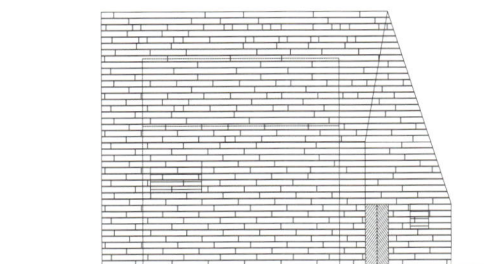

西立面 west elevation

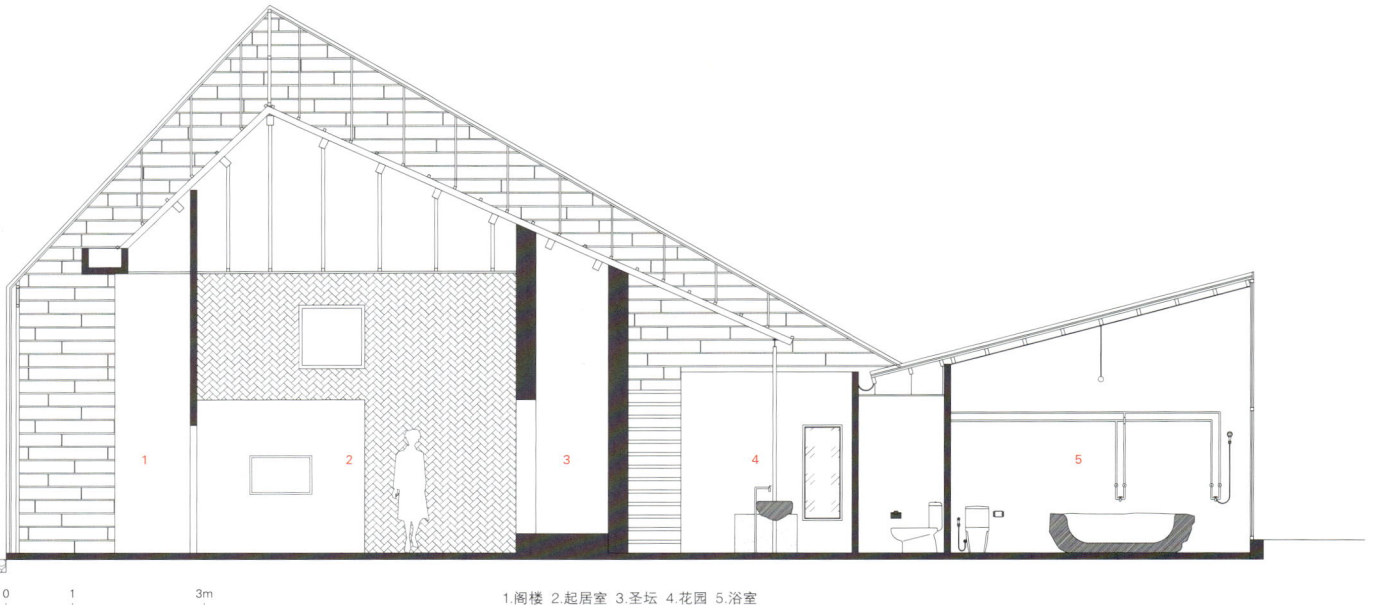

1.阁楼 2.起居室 3.圣坛 4.花园 5.浴室
1. loft 2. living room 3. altar 4. garden 5. bathroom
A-A' 剖面图 section A-A'

项目名称：FA House
地点：Lam Dong, Vietnam
建筑师：atelier tho.A
用地面积：1,100m²
总建筑面积：160m²
有效楼层面积：195m²
设计时间：2014
施工时间：2014
竣工时间：2014
摄影师：©Quangtran (courtesy of the architect)

城市环境内的密集居所

Density in

特殊场地内住宅的设计策略
Design Strategies for Site-Specific Housing

　　我们现今居住的城市人口越来越密集，因为变化的现代生活方式使得人们靠近活跃的城市中心区生活，在这里，他们既可以参与丰富的文化活动，又可以享受公共设施带来的便利生活。密集的城市住宅一方面给人们带来了方便，然而另一方面，却可能造成多样化人口共同生活的不便。当今这种快节奏的生活方式，让我们很难与邻里和周围建立联系，因此人群的多样性，并没有为城市生活带来优势，反而对人们的隐私与安全造成威胁。近期，就这一问题，人们开发了中等密度住宅项目，利用各种策略让单独的住宅单元之间（也包括与周围环境之间）建立起联系。建立联系的方式有三种：赋予建筑一个强烈且具体的视觉效果，该视觉效果起源于场地的形状，并且与之建立对话；为居民之间的社交提供公共场所；利用周边的公共空间将私人区域和这些公共场所连接起来。

Our cities are becoming increasingly dense, as changing contemporary lifestyles attract people to live close to active urban centres, where they can benefit from cultural activities and the conveniences of public services. With the conveniences of dense urban dwellings, however, comes the potential inconvenience of living among a diverse population. As today's hectic lifestyles make it difficult to establish links with our neighbourhoods and surroundings, this diversity risks becoming a threat to our privacy and security, instead of an advantage of urban life. Related to this challenge, recent medium-density housing projects are developing various strategies for establishing connections with the surrounding context, and between the individual dwelling units within. These connections are achieved in three ways: by giving the project a strong and specific visual identity that derives from and dialogues with the formal qualities of the site, by providing communal places for social exchange between residents, and by connecting these internal and communal areas with the surrounding public space.

SL11024住宅_SL11024/Lorcan O'Herlihy Architects
濑户某公司住宅_Seto Company Housing/Mount Fuji Architects Studio
Brugg镇的阶梯式住宅_Terraced Housing in Brugg/Ken Architekten
Tappen住宅_Tappen Housing/Joliark
Nantes地区某住宅_Housing in Nantes/Antonini Darmon
Grand-Pré某住宅_Housing of Grand-Pré/Luscher Architectes SA

城市环境内的密集居所：特殊场地内住宅的设计策略_Density in Context: Design Strategies for Site-Specific Housing/Isabel Potworowski

Context

密集型居所的优势

针对当下城市不断发展扩大的问题，针对城市人口密集化而开发合适策略成为一个挑战，包括城市规划的各个方面，如城市基础结构、公共设施和住房的策略设计。一般来说，解决住房密集化问题主要集中在翻新过时的城市结构，限制城市对外扩张，由此密集化城市周边区域，以保持城市与自然的连接，同时发展混合型和集约型的城市布局[1]。正是不断增值的土地和密集型生活的可持续性推动了城市的发展变化。城市的密集化优化了土地利用，从而缓解了耕地面积不足造成的压力，同时缩短了人们的出行距离，突出了公共交通工具的优势。

居住在人口稠密的市中心，不仅推动了可持续性的发展，同时还引领了这个时代所形成的新型的生活方式。传统的核心家庭形式逐渐消失，取而代之的形式的多种多样：单身生活、丁克家庭、单亲家庭以及新一轮的老龄化家庭和移民家庭。显然，这些群体更享受城市生活：便利的公共交通和服务、丰富的文化网络和为残疾人提供的舒适生活[2]。

直面多样化

虽然城市密集化设计带来生活的便利，却也隐藏着多样化带来的不便。现今的社会，多种多样的生活方式和文化形式并存，近期欧洲的移民使社会变得空前多元化。群居生活会引发关于隐私和安全方面的问题，如果这些问题得不到解决，会引发更严重的后果。"如今，有关社区陌生人之间的交往的问题，即公共领域，是通过彼此间保持距离来达到和谐共存的目的的。" Richard Sennett解释道，"也就是说，人们选择保持安全距离，与他人的冷漠相处换取和平共处。"这种冷漠的方

The Advantages of Density

Developing appropriate strategies for densification is a challenge that today's growing cities are increasingly being faced with, the one that touches upon all aspects of urban planning, from policy-making to the design of infrastructure, public amenities and housing. In general, densification efforts in housing developments have been concentrated on the renewal of worn-out urban fabric, on densifying urban peripheries to preserve the connection with natural environments by limiting sprawl, and on developing mixed, well-integrated organisations[1]. These developments are being propelled by the rising cost of land, and by the sustainability of living closer together. Density optimises land use, lessens the pressure on agricultural land, reduces the need to travel, and makes public transportation more profitable. Beyond sustainability, though, living in close proximity to a dense urban centre appeals to contemporary lifestyles. As the classical nuclear family disappears, diverse subgroups are appearing such as singles, "dinkies" (double income no kids), one-parent families, and new waves of elderly and immigrants. Almost all of these groups prefer an urban environment for the associated benefits of public transportation and services, cultural networks, and comfort for the disabled[2].

Confronting Diversity

With the convenience of density, however, comes the potential inconvenience of confronting diversity. Our pluralist society juxtaposes various lifestyles and cultures, which are becoming even more in the case with the recent European migrant crisis. Living together can provoke questions concerning privacy and security, which can cause problems if left unresolved. "About the sociability of living with strangers," explains Richard Sennett, "the mark of the civic realm now is the mutual accommodation through dissociation. That means the truce of letting one another alone, the peace of mutual indifference." While indifference may be convenient, however, it "spells the end of citizenship practices which require understanding of divergent interests," and marks "a loss of

Brugg镇的阶梯式住宅，瑞士
Terraced Housing in Brugg, Switzerland

式可能是便利的，但与此同时，它也"阻止了不同兴趣爱好的市民之间的了解"，标志着"市民对他人的基本好奇心的缺失"³。这种冷漠的态度导致美国封闭式社区的日益涌现。2001年的调查显示，新建社区中有80％为封闭式社区⁴。而911恐怖袭击更提高了许多市民的安全警惕和反恐意识。然而，对"他人"的冷漠和零容忍并不只出现在美国的社会环境中，这种心态在当今纷繁复杂的社会中几乎无处不在。如今人们的忙碌和快节奏的生活方式，基本不可能使邻里之间建立联系。人们频繁更换工作，随之而来的便是生活环境的不断变化。以欧洲人为例，欧洲人平均每三年到五年就会搬一次家⁵。在没有亲近的邻里间的关系的前提下，我们的城市如何向互通的社会迈进一步？建筑设计又该如何面对这一挑战？

更中性化，更个性化

随着城市的密集化以及迎合不同人群的需求，两种不同的趋势展现在人们面前。一方面，由于生活方式各有不同，转售比率不断提高，建筑师对顾客越来越莫衷一是，所以房屋的装潢越来越中性化，越来越灵活⁶。相比现代公寓原型，如今的住宅功能有所"退化"，都以可选择的实用元素为主，给人们设定自由分隔的单一空间概念⁷。

而另一方面，住宅开发愈加个性鲜明。生活在这个全球化的大众社会中，每个个体都在努力寻求自己的身份定位，这个定位中一个重要的方面就是自己的住宅场地⁸。现今的建筑设计都有很强烈的，甚至是标志性的视觉特色，来表现其与周围环境相联系的方式。有趣的是，场地不论是乡村周边还是翻修的工厂区，建筑师都会用建筑周围的环境来突出建筑本身的特色。

simple human curiosity about The Other."³ The consequences of such an attitude are reflected in, for instance, the increasing number of gated communities in the United States. A survey carried out in 2001 indicates that eighty percent of new residential estates are being developed within an enclosure.⁴ The 9/11 attacks have certainly contributed to the sense of security and fear of many groups of citizens. That being said, though, the risk of indifference and intolerance of "The Other" is not limited to the American context. It is a risk in every city, and today's hectic and dynamic lifestyles make it increasingly difficult to build relations with one's neighbours and surroundings. People are changing their jobs more often, so is their living environment. Europeans, for instance, move on average every three to five years.⁵ In the absence of strong links with one's neighbours and surroundings, how can our cities develop and move forward on a social level? What role can the design of housing developments play in addressing this challenge?

More Neutral and More Specific

In parallel with urban densification and with the need to accommodate a diverse population, two trends have been emerging. On the one hand, dwellings have increasingly neutral and flexible interiors. Because of the diversity of lifestyles and increasing turnover rates, architects no longer know for whom they are building.⁶ As opposed to the proto-typical modernist apartment, today's dwellings are "de-programmed," offering freely divisible single-space concepts with optional placements of utility elements.⁷ On the other hand, housing developments are becoming more specific. In a global mass society, people search desperately for identity, an important factor of which is the dwelling place.⁸ Accordingly, recent housing projects have a strong – even iconic – visual character, often expressing how they connect with their context. It is interesting to observe that, in various sites ranging from village peripheries to regenerated industrial areas, architects give their projects a strong identity

瀬户某公司住宅，福山，日本
Seto Company Housing in Fukuyama, Japan

Luscher建筑师事务所在Grand-Pré建造的SA住宅，位于瑞士的一个历史悠久的小镇Crans-près-Céligny边缘的农耕地周围。建筑师形容这座建筑的特色为："用类似于立体派绘画设计来诠释村落的实质，"并用石灰石材料将其实体化。五栋单元楼体相互交错坐落，反映出历史小镇的空间格局、视觉效果和人行道的通透性。每一个体量都由更小的立体体量组成，以此来尊重现有周围环境的尺度。单元楼间的开放空间依次由街道附近的石板、木板和路边草坪向周边更茂盛的植被转化，以展现公共和私人空间、城镇范围和农田范围之间的过渡性。

Ken建筑事务所的Brugg镇阶梯式住宅也位于一处类似的环境中，即瑞士的一个丘陵地貌的小镇——Brugg的边缘。建筑师利用自然地形来建立这个项目的特色。这是一个有16个住宅单元的阶梯式连体公寓，其中央楼梯的两侧各有八个单元。各个单元又融为一个封闭的整

体体量，整个楼体嵌入一侧的Brugg山体中。楼体的大型规模、倾斜角度、微小细节和天然石色都成为Brugg自然景观中的一部分。

富士山建筑师工作室设计的濑户公司船厂工人公寓的背景，与Brugg镇的阶梯式住宅的背景具有相似性。这座工人公寓位于日本濑户内海海滨城市中主要的工业城市之———福山附近的村庄，其设计应用了解决环境方案中的第三种策略。这个村庄位于直接与海相接的山上，由于缺少天然的平坦地形，人们没有可以聚集在一起的公共空间。为此，建筑师在山体中嵌入了一个宽阔的三层悬挑建筑。人们可以在建筑屋顶的大型社区公共广场中欣赏海景。一个高高的体量平衡了悬挑体量的承重力，使整座建筑呈现L形。人们可以通过屋顶公共广场上的两个露天庭院进入下面的公寓，实现了私人流线和公共空间的完美连接。

that integrates them with the surroundings.
Luscher Architectes SA's Housing of Grand-Pré is situated on the border of the small historic Swiss town Crans-près-Céligny and the surrounding agricultural land. The architects describe the project's identity as a "quasi-Cubist pictorial interpretation of the essence of the village", materialized in travertine stone. The five volumes of dwelling units are arranged on the plot in a staggered configuration, mirroring the spatial pattern, visual and pedestrian permeability of the historic town. Each volume is a "Cubist" composition of smaller volumes, respecting the scale of the surroundings. The network of open spaces between the units changes gradually from stone, wood and grass near the street to more dense vegetation, marking the transition from public to private and from town to farmland. Ken Architekten's Terraced Housing in Brugg project is located in a similar setting, on the periphery of Brugg, a Swiss town with a hilly landscape. Here, the architects have looked to the natural topography for establishing the project's identity.

Sixteen terraced condominium apartments, organized as eight units on either side of a central stairway, are fused into a closed overall volume that embeds itself into the side of the Bruggerberg hill. The volume's large scale, oblique angles, minimal detail and natural stone colour give the impression that it is part of the natural landscape.
Operating in a comparable setting, Mount Fuji Architects Studio's Seto Company Housing for shipyard workers in a Japanese village near Fukuyama, one of several major industrial cities on the coastland area of the Seto Inland Sea, presents a third strategy for addressing the context. The village is situated on hills that descend directly to the sea, and lacked a public space for gatherings because of the naturally limited flat land. In response, the architects have embedded a wide, three-story cantilevering volume in the hill, its roof providing a large public plaza for the community with a view to the sea. A taller housing volume that acts as a counter-weight completes the L-shaped building. Two light courts in the rooftop

照片提供：©Alexandre Wasilewski

Nantes地区某住宅，法国
Housing in Nantes, France

　　Joliark事务所设计的Trappen住宅坐落在一座人口更加密集的城市中。斯德哥尔摩郊外的一个老工业区——玛丽霍尔已改建成为一个半封闭的城区，这座建筑就是改建工程的一部分。它有两个主楼体，共44个住宅，形成了这个城区的一部分，其中还建有一个中心公共花园。很大程度上，这座建筑的特点和个性设计都源于其独特的建筑材料：镀锌钢板的外墙犹如"防弹衣"一样包裹着整座建筑。而在临街的西立面，双高的阳台却使用了木材来包裹，来形成反差。混凝土墙体和墙板形成简单而大胆的网格，穿插于阳台表面，将独立的单元更加清晰地呈现在路人面前。而在体块的内部，人们可在花园和四周的走廊进行社交活动。

　　同样位于老工业区的建筑还有Antonini Darmon设计的位于Île de Nantes西部的标志性混合式社交住宅。该住宅名为Oiseau des Iles，即

"岛屿上的鸟"，其白钢包裹的十层住宅塔楼置于整体由木板包裹的基座上，基座内设停车场、商圈和六个住宅单元。这两个因素所带来的双重性赋予了该设计强烈的个性特色，使它在周围多样化的工业建筑群（重新规划的工业建筑建筑、写字楼和公共场所）中脱颖而出。设有商业区的基座与街道相连，楼体的立面也利用墙体、上空和穿孔的围屏的形式与环境进行视觉连接。同样，基座的带有棱角的屋顶表面还形成了一处悬空的甲板空间，该空间与住宅楼一层入口处的绿植区域共同为居民提供可以聚会的场所。

　　Lorcan O'Herlihy建筑师事务所设计的SL11024师生公寓综合楼展现了生动的公共空间。它位于人口密集的洛杉矶韦斯特伍德社区，在加州大学洛杉矶分校区附近。整座建筑依山而建，景观式屋顶露台直接从街道延伸至屋顶。这些聚会场所和流通空间是从白钢体量中切割出

plaza give access to the apartments below, making a strong connection between private circulation and public space. Joliark's Trappen Housing is situated in a more dense urban setting. It is part of the ongoing regeneration of Mariehäll, a former industrial site outside Stockholm that is being redeveloped as semi-enclosed urban blocks. The project comprises 44 dwellings organized in two volumes that form part of an urban block, and includes a central communal garden. It receives its identity and character largely from its material expression: a galvanized steel sheet facade wraps around the structure like a "protective cloak". On its western elevation, facing the street, double-height balcony compartments have a contrasting materialization of wood. The balcony compartments are articulated by a simple and bold grid of concrete walls and slabs, which makes the individual units more legible to the passer-by. In the interior of the block, the garden and surrounding access galleries provide opportunities for social interaction.

Likewise working in a former industrial area, Antonini Darmon have designed an iconic form for their mixed-use social housing project on the Western tip of Île de Nantes. Titled Oiseau des Iles, "Bird of the Islands", the 10-story residential tower clad in white steel rests on a monolithic, timber-clad plinth that contains parking, six dwelling units and commercial program. The duality of these two elements gives the project a strong identity amidst the surrounding heterogenous ensemble of repurposed industrial buildings, offices and public functions. The commercial plinth establishes a connection with the street, and the tower's facade visually connects with the surroundings through the pattern of solids, voids and perforated screens. As well, the plinth's angular roof surface creates an elevated deck topography which, together with the planted areas that frame the tower's first-floor entrance, provides a meeting place for residents.
The identity of Lorcan O'Herlihy Architects' SL11024 student and faculty housing complex derives from the vivid expres-

SL11024住宅，洛杉矶，加利福尼亚州，美国
SL11024 in Los Angeles, CA, United States

来的缺口，呈现为亮绿色混凝土表面，建筑立面覆层系统利用坚固的穿孔白色金属肋板（展现了建筑内部结构和功能的活力）连接了住宅群和街道。

　　以上这些住宅设计都具有强烈的特殊形式特征，这源于其周围独特的城市或周围自然环境特点，也得益于它们内部对于私人单元和公共空间的独特规划。从一定意义来讲，它们中的每一个都是一块飞地，用自己独特的个性来诠释它们周围的城市和公共空间。同时，公共空间也诠释着每一个独立单元之间的联系。这样的格局表明了当代城市发展的方向，就是要形成这种飞地的组合体⁹。住宅区不再是封闭式社区，而是彼此相连甚至与公共空间相连的独立个体，鼓励独立的人们利用公共空间与"他人"交流。

1. Javier Mozas, Aurora Fernández Per, "Collective Housing. 10 stamps", *Density: New Collective Housing*, Vitoria-Gasteiz: a+t Ediciones, 2004, p.207
2. Atelier Kempe Thill, "Specific Neutrality", *Density: New Collective Housing*, Vitoria-Gasteiz: a+t Ediciones, 2004, p.138
3. Richard Sennet, cited in Javier Mozas, p.211
4. Javier Mozas, p.211
5. Atelier Kempe Thill, pp.138~143
6. Idem, p.138
7. Ernst Hubeli's contribution in Reinhard Gieselmann, "Historical Development of Housing Plans", *Floor Plan Manual: Housing*, Ed. Friederike Schneider, Basel: Birkhäuser, 2004, p.32
8. Atelier Kempe Thill, p.142
9. Dick van Gameren, Pierijn van der Putt, "Fragments of an Ideal City", *Delft Architectural Studies on Housing (DASH)*, 2011, pp.5~9

sion of its communal spaces. Located in the dense established Los Angeles neighbourhood of Westwood, nearby the UCLA campus, the building responds to the hillside topography with landscaped roof terraces that step back continuously from street level to the roof. These gathering spaces, as well as the circulation voids, are cuts made in the white steel volume, revealing bright green concrete surfaces. The facade cladding system also establishes a connection between the housing complex and the street by solid, perforated and ribbed white metal panels that project the dynamism of the internal programming and functions.

These housing projects all have a strong and specific formal character, which derives from the particular urban or natural character of their surroundings, and from the interior organization of individual units and common spaces. In a certain sense, each one is an enclave, with its own identity that expresses its interpretation of the surrounding city, and common spaces that suggest a particular relation between the dwelling units within. This pattern suggests a direction for the contemporary city as a collage of such enclaves;⁹ not as gated communities, but as distinct entities connected to each other and to the public space in between, encouraging social exchange through an openness with "Others". Isabel Potworowski

SL11024住宅

Lorcan O'Herlihy Architects

这个项目综合体利用其建筑材料和造型,将自身与其周围富有历史气息的场地和具有挑战性的山地地形完美融合,创造出了一个城市发展新模式,来完善学术社区的建设。住宅群位于洛杉矶韦斯特伍德小区,对面是Richard Neutra设计的Strathmore公寓。建筑项目共有31个单元楼,还设有娱乐设施,同时在加州大学洛杉矶分校区的周边社区设置了可以提供紧急住房和聚会社交的空间,以向其周边的优质社区表达敬意。通过与客户和社区管理人员的沟通,建筑师提出了一个方案,可以同时满足周边原有社区住户的需求、本社区客户的大量功能要求以及针对独特的楔形地形的建筑要求。

整座建筑被分为两个体量,以便建筑师可以将现有的景观和整体体量有效地整合起来,保证公寓内的交叉通风,同时确保建筑内有清晰的流线。整个建筑体量沿着街道的自然坡度而逐渐下移,其最低点横穿Neutra设计的现代主义地标处,从而达到呼应并放大其周边备受瞩目的场地的战略。

建筑师不断努力,想要在密集化发展的城市中加入一些绿色空间,因此,这座建筑也结合了阶梯式景观屋顶露台,为学生营造迷人的室外环境。建筑师们利用不断后退的建筑造型,构建出从每个楼层都可以进入屋顶平台或庭院的通道,为学生们的聚会和社交提供了重要的场所。这种从街道一直延续到屋顶的阶梯式社交空间共同成就了强有力的城市建筑造型,它为建筑内部和面向周围社区的外部之间建立了多层次的联系。

建筑师将坚固的穿孔白色金属肋条板作为建筑的外围护结构,为建筑造型、结构和功能嵌入活力。八层绿色调的水泥板从建筑底部到顶部形成了渐变的效果,而深浅不一的绿色也同时划分了建筑的不同区域,以形成流通空间和社交空间。建筑覆层系统运用动画的光影效果打破了墙面的规模束缚,模糊了人行道和建筑群的界限。

0　10　　30m

这个项目是建筑师不断努力解决和改善城市环境的一部分。这里，LOHA事务所运用独特的山地地形，打造出一座多面的建筑，这将成为韦斯特伍德在城市不断密集化发展过程中的财富。

SL11024

Through its materiality and form, design for the housing complex seamlessly engages its historically sensitive site and challenging hillside topography and creates a new model for urban development that enriches an academic community. Sited opposite Richard Neutra's Strathmore Apartments in the Westwood neighborhood of Los Angeles, this student and faculty housing complex of 31 units and recreational amenities pays homage to its preeminent neighbor while providing the community with much-needed housing and gathering spaces on the border of University of California, Los Angeles(UCLA)'s campus. Following conversations with the client and community activists, the architects developed a scheme that responds to the considerations of the established neighborhood, the client's extensive program requirements, and the unique, wedge-shaped site.

By splitting the building into two volumes, the architects were able to efficiently integrate the structure's overall massing with the existing landscape, provide wind cross-ventilation to the apartments, and define a clear circulation path through the property. The building's volumes shift downward along the street's natural incline and reach their lowest height directly across from Neutra's modernist landmark, thereby echoing and amplifying the enlightened site strategies of this high-profile neighbor.

项目名称：SL11024
地点：Los Angeles, CA
建筑师：Lorcan O'Herlihy Architects
主要建筑师：Lorcan O'Herlihy, FAIA
设计团队：Donnie Schmidt _ project manager, Damian LeMons, Lilit Ustayan, Chris Faulhammer, Ian Dickenson, Abel Garcia
功能：Off-campus housing catering to university community, fitness center, yoga studio, business center, resident lounges
用地面积：1,938.42m²
总建筑面积：3,901.93m²
有效楼层面积：5,110.67m²
设计时间：2012
施工时间：2014
竣工时间：2015
摄影师：©Iwan Baan (courtesy of the architect)

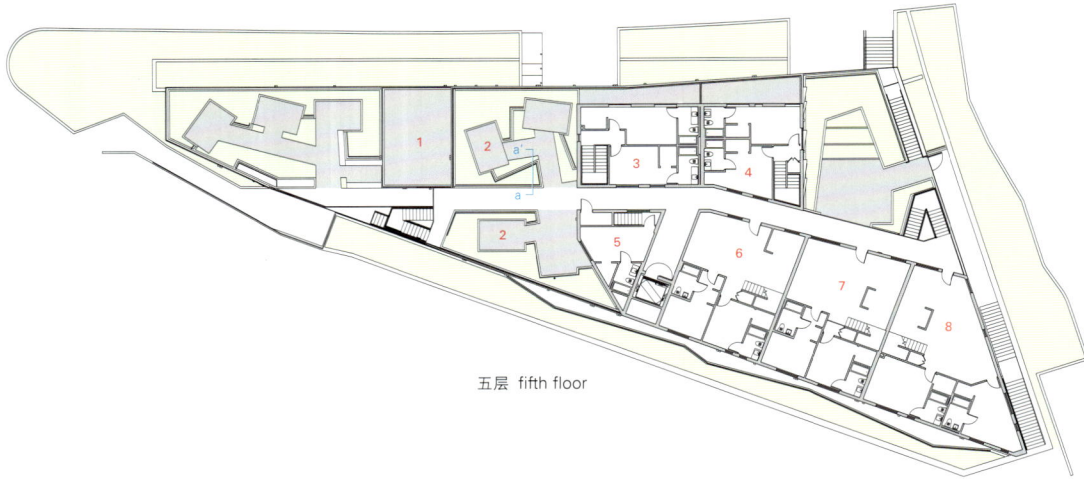

五层 fifth floor

1.四层屋顶平台
2.屋顶平台
3.Q单元
4.R单元
5.S单元
6.T单元
7.U单元
8.V单元

1. 4th floor roof deck
2. roof deck
3. unit Q
4. unit R
5. unit S
6. unit T
7. unit U
8. unit V

三层 third floor

1.屋顶平台
2.I-2单元
3.O单元
4.P单元
5.L单元
6.M单元
7.D单元
8.N单元

1. roof deck
2. unit I-2
3. unit O
4. unit P
5. unit L
6. unit M
7. unit D
8. unit N

一层 first floor

1.庭院
2.大厅
3.办公室
4.休息室
5.公共区
6.会议室
7.健身中心
8.A单元
9.B单元
10.C单元
11.D单元
12.E单元

1. courtyard
2. lobby
3. office
4. lounge
5. common room
6. conference room
7. fitness center
8. unit A
9. unit B
10. unit C
11. unit D
12. unit E

0   5   10m

Continuing architect's efforts to add green spaces into dense, urban developments, the building incorporates landscaped roof terraces at various levels, creating inviting outdoor areas for students. By stepping back the form, the architects created access from each floor level to either a roof deck or a courtyard space, providing vital gathering and community-building venues. This continuous terracing of communal space from the street level to the roof results in a powerful urban and architectural gesture that offers a multilayer connection both within the building and out towards the surrounding community.

The architect's material choices for the building envelope of solid, perforated, and ribbed white metal panels project the vibrancy embedded within the project's form, programming, and functions. Cement board painted in eight green-hued layers creates a gradient effect that grounds the building's base and lightens as the structure meets the sky. The green delineates cuts in the building's form, indicating voids for circulation and gathering spaces. The cladding system blurs the boundary between the sidewalk and the building mass, creating an animated effect of rippling shadow and light that breaks up the scale of the wall surfaces.

This project is a part of the architect's ongoing commitment to address and elevate challenging urban conditions. Here, LOHA embraced the unique topography of a hillside site to craft a multifaceted building that is an asset to the increasingly dense and dynamic urban fabric of Westwood.

1. 24 ga. sht. mtl. coping (painted)
2. self-ad waterproof underlayment membrane type
3. 42" h. mtl. rail system
4. 18 ga steel studs @16" oc
5. metal panel over weather barrier type

6. 1/2" plywood sheathing
7. 24 ga ss flashing over self-ad wpr. membrane. underlayment
8. 5/8" wide stl plate @48" oc.
9. 3"x5" steel angled clip @48" oc to fasten guardrail to rim beam
10. 2 layers 5/8" DensGlass sheathing

11. rough carpentry
12. sched. window
13. sched. floor assembly
14. 3/4" plywood o/2x sleepers type
15. earth type
16. hot fluid applied water proof membrane type
17. 1/4 dense-deck
18. foam block
19. fiber cement board panels over weather barrier type
20. 2x wood decking over 2x sleeper o/ neoprene pads type
21. 18 ga ss brake shaped planter curb
22. provide 1" dia weep @ 8" oc type
23. 24 ga ss flashing o/ self adhesive waterproofing
    membrane underlayment type
24. root barrier type
25. drain mat
26. 1/2" gyp. bd.

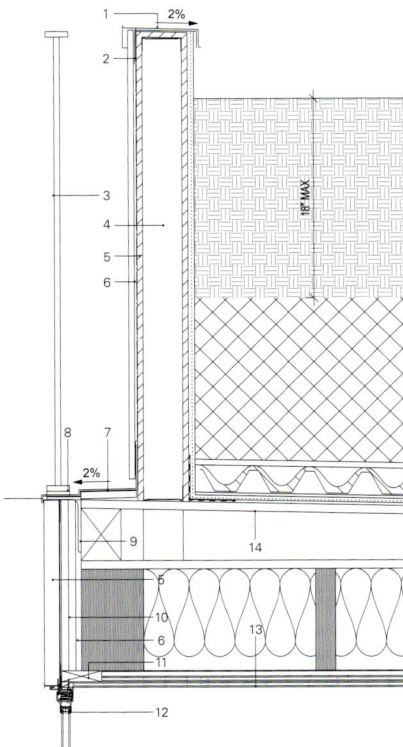

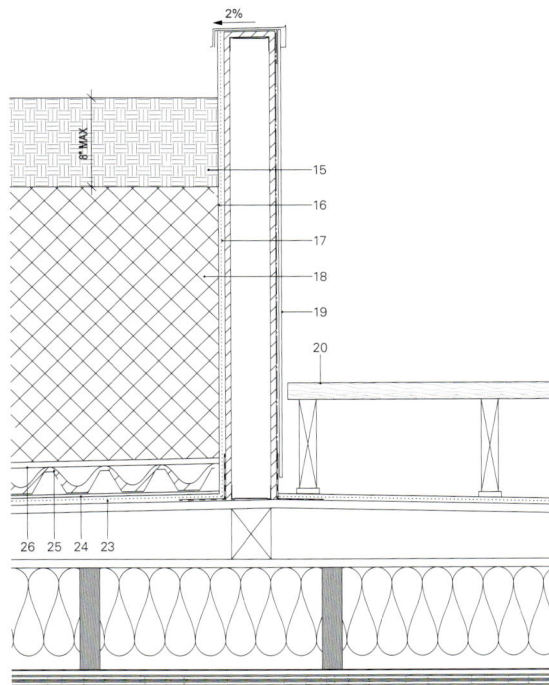

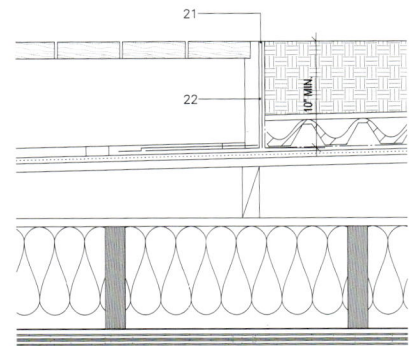

a-a' 剖面图 section a-a'

北立面 north elevation

0 2 5m

西立面 west elevation

1.A单元	8.C单元
2.大厅	9.B单元
3.M单元	10.N单元
4.屋顶平台	11.D单元
5.庭院	12.R单元
6.E单元	13.Q单元
7.D单元	

1. unit A	8. unit C
2. lobby	9. unit B
3. unit M	10. unit N
4. roof deck	11. unit D
5. courtyard	12. unit R
6. unit E	13. unit Q
7. unit D	

A–A' 剖面图
section A-A'

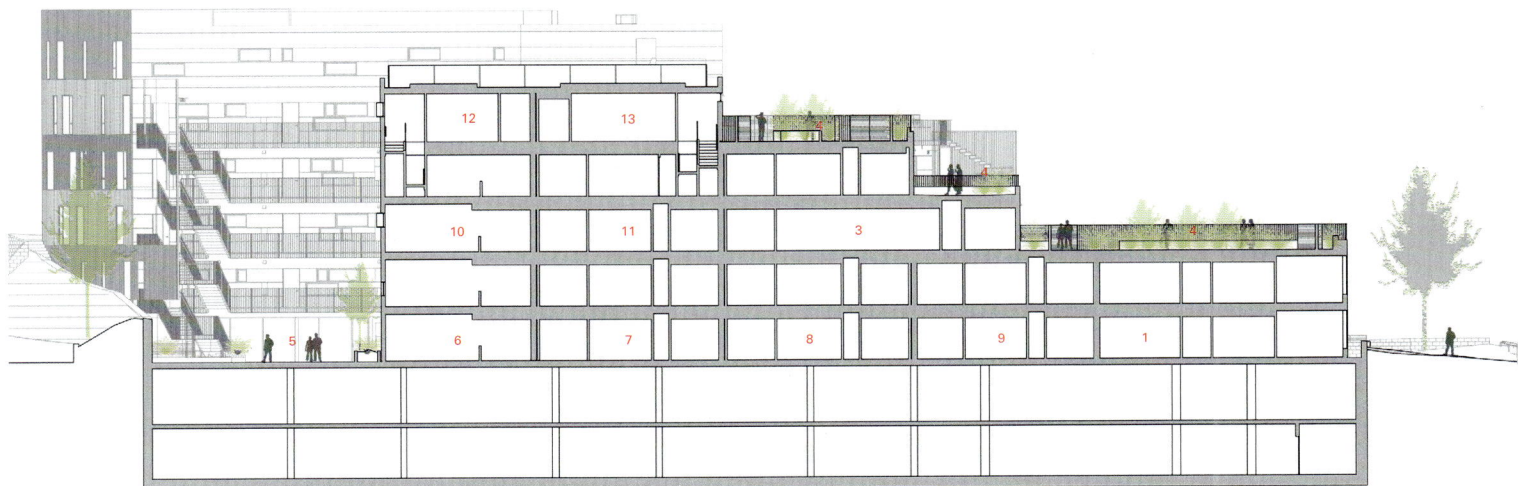

B–B' 剖面图 section B-B'

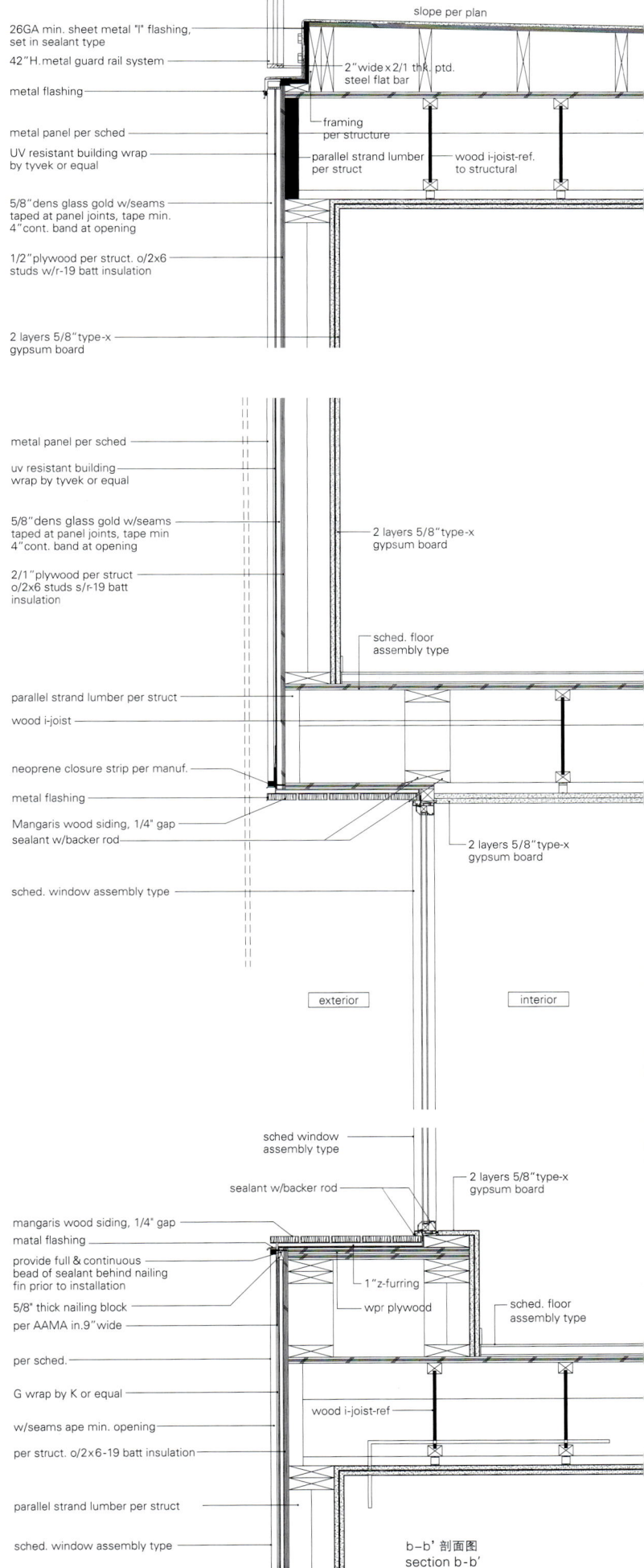

slope per plan

26GA min. sheet metal "I" flashing, set in sealant type

42"H.metal guard rail system

metal flashing

2"wide x 2/1 thk. ptd. steel flat bar

metal panel per sched

framing per structure

UV resistant building wrap by tyvek or equal

parallel strand lumber per struct

wood i-joist-ref. to structural

5/8"dens glass gold w/seams taped at panel joints, tape min. 4"cont. band at opening

1/2"plywood per struct. o/2x6 studs w/r-19 batt insulation

2 layers 5/8"type-x gypsum board

metal panel per sched

uv resistant building wrap by tyvek or equal

5/8"dens glass gold w/seams taped at panel joints, tape min 4"cont. band at opening

2 layers 5/8"type-x gypsum board

2/1"plywood per struct o/2x6 studs s/r-19 batt insulation

sched. floor assembly type

parallel strand lumber per struct

wood i-joist

neoprene closure strip per manuf.

metal flashing

Mangaris wood siding, 1/4" gap

sealant w/backer rod

2 layers 5/8"type-x gypsum board

sched. window assembly type

exterior

interior

sched window assembly type

2 layers 5/8"type-x gypsum board

sealant w/backer rod

mangaris wood siding, 1/4" gap

matal flashing

provide full & continuous bead of sealant behind nailing fin prior to installation

1"z-furring

wpr plywood

5/8" thick nailing block per AAMA in.9"wide

sched. floor assembly type

per sched.

G wrap by K or equal

wood i-joist-ref

w/seams ape min. opening

per struct. o/2x6-19 batt insulation

parallel strand lumber per struct

sched. window assembly type

b-b' 剖面图
section b-b'

濑户某公司住宅

Mount Fuji Architects Studio

迎着美丽且平静、小岛漂浮的濑户内海，一个造船企业员工公寓依悬崖而建。这个项目被寄予厚望，以振兴当地的工业城市，同时也是37个家庭的温暖舒适住所。屋顶从山坡延伸出来，使屋顶平台形成了一个在丘陵地势中少见的宽敞的公共交流空间。小镇就建在陡峭的山坡上，山顶上是村落森林，山脚就是一片大海，这就使当地居民缺乏平地的公共空间来聚集或举办活动。而我们试图利用这个坡度建造出一个公共空间。

我们设计的是一个三层的住宅结构，其中一半楼体是悬于悬崖外的。拥有绝佳海景的屋顶是对大众开放的屋顶广场，从那里通过宽敞的楼梯可以直接进入到北部道路。在悬空楼体的底部，是另一个封闭式的公共空间，可以让人们在里面避雨。

两个与中国窑洞设计相仿的露天庭院是进入到居住区的通道，同时也保证了原本昏暗的中央走廊和建筑中的每个房间有足够的日照和通风。

建筑造型就像是一艘等待下水仪式的船，但这不只是一个设计。这也是一个象征，象征着建筑设计的自主性和城市、环境、结构、经济等因素对建筑的约束之间达到了真正的平衡，用船的造型表现出了"平衡"的特性。

我们在基桩和悬崖之间保留了足够的距离，以保证在建造宽阔的屋顶公共空间时，不给悬崖造成过大压力。因此我们采用悬臂拼装法。用"连续结构墙"做出三层楼体的横梁，并根据压力的变化将"预应力混凝土梁"或弯曲或伸展地穿插在其中，由此形成一个可行的方案。

将预应力混凝土梁进行弯曲的方法理论上是可行的，但由于实践过程中困难重重，因此这个方法之前尚未在现实世界中应用，也就是说，这是领先于世界的技术试验。随着对这一技术的成功突破，这个新的结构一定会投入日常实践中。

其实，美丽的海景住宅塔楼也就是一个平衡配重的悬臂式楼体。

A shipbuilder's company housing sits on a cliff viewing beautiful and calm Seto Inland Sea, where small islands float. This project is expected to revitalize a local industrial city while serving as comfortable accommodations for 37 families. On the roof that extends from ground level uphill, the rooftop plaza, a rare public space in the hilly town spreads out.
The town is posed on a steep hill. On the hilltop is the village forests, and downhill, the ocean. Lack of flatland deprives the residents of enough public space to gather and hold events. Taking advantage of slope, we tried to create a public space architecturally.

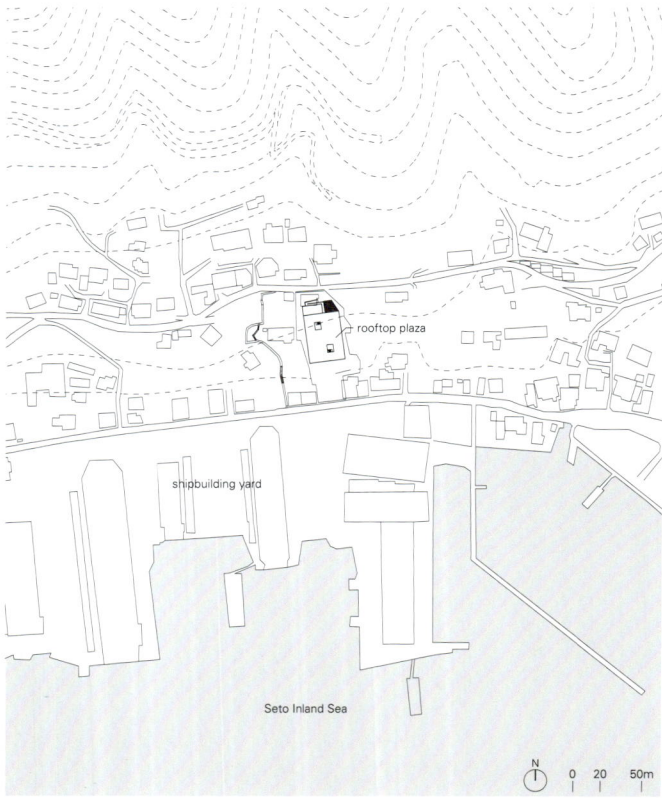

rooftop plaza

shipbuilding yard

Seto Inland Sea

N 0 20 50m

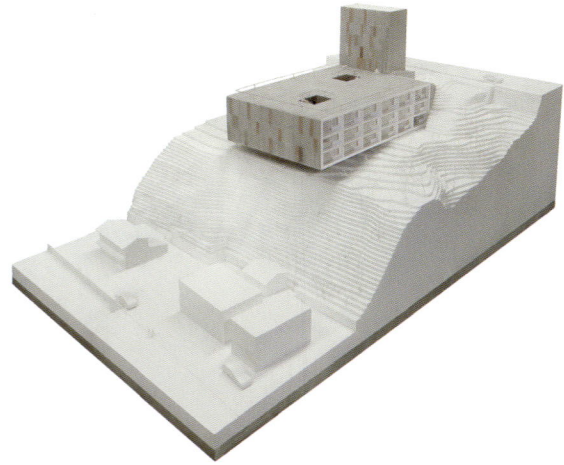

南立面 south elevation

东立面 east elevation

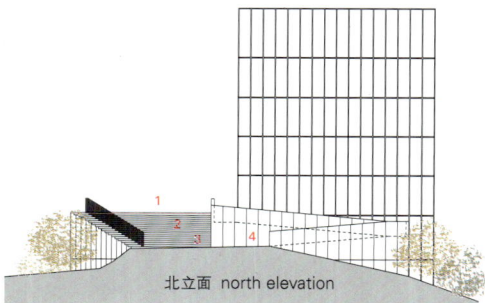

北立面 north elevation

0 2 5m

1.屋顶广场　1. rooftop plaza
2.大楼梯　　2. grand staircase
3.入口通道　3. entrance approach
4.北侧道路　4. north road

屋顶 roof

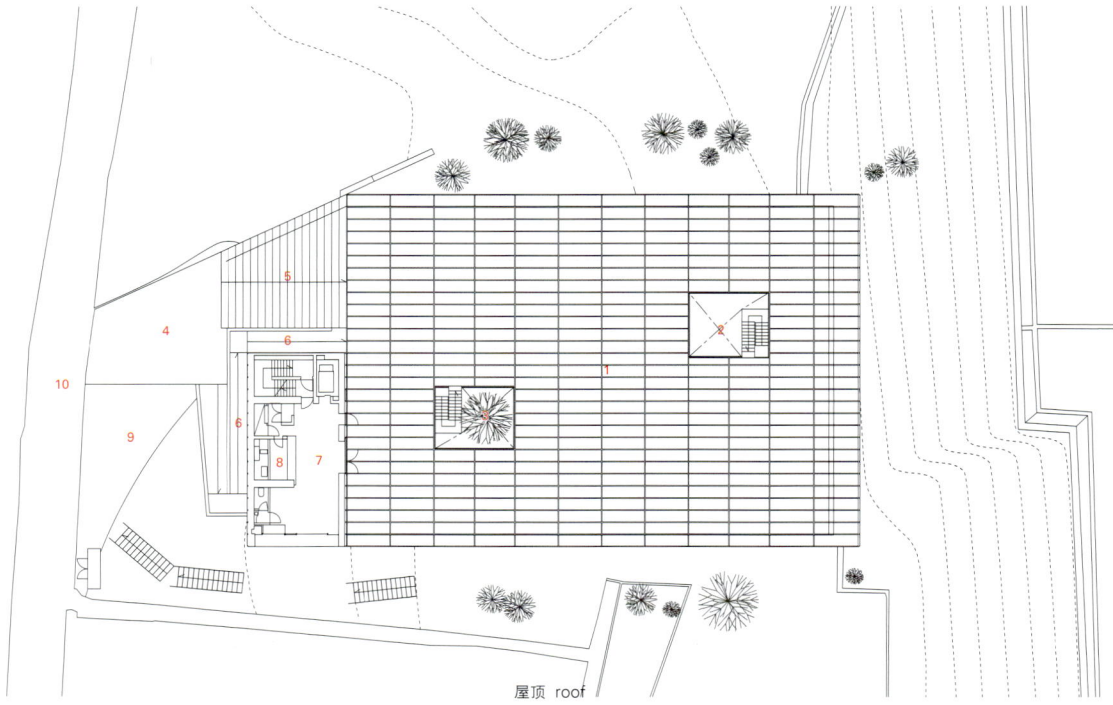

1.屋顶广场
2.露台庭院1
3.露台庭院2
4.入口通道
5.大楼梯
6.斜坡
7.休息室
8.酒吧
9.停车场
10.北侧道路
11.露台
12.储藏室

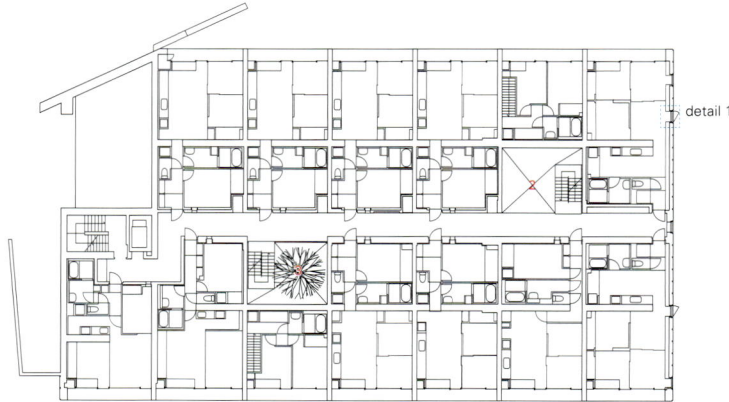

1. rooftop plaza
2. light court 1
3. light court 2
4. entrance approach
5. grand staircase
6. slope
7. lounge
8. bar
9. parking
10. north road
11. terrace
12. storage

detail 1

三层 third floor

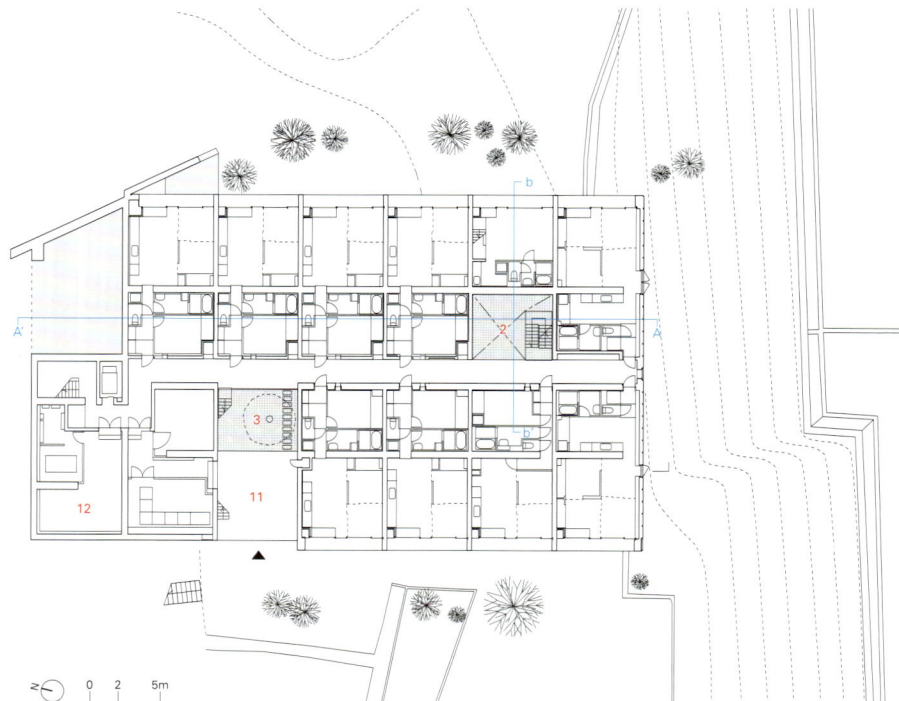

一层 first floor

N   0 2   5m

130
70 60
60

2,480

110
50 60
3 25 4 28
8 8 10

60

70 60
130

a–a' 剖面图
section a-a'

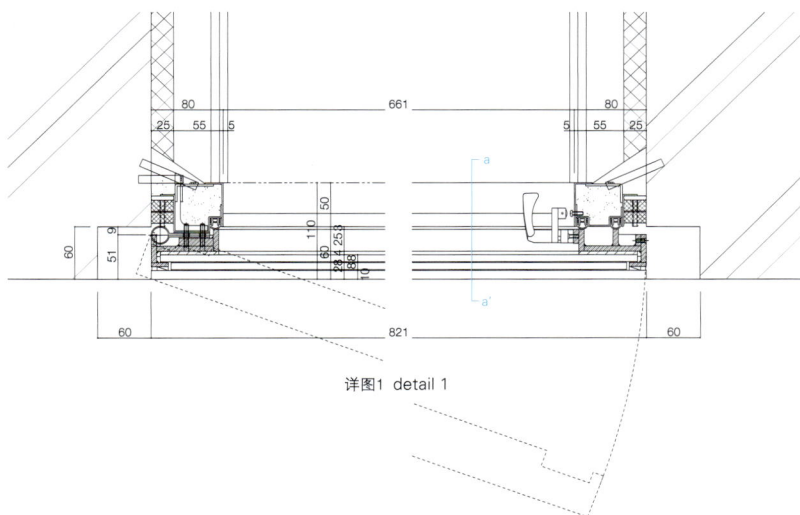

80　661　80
25 55 5　5 55 25

a
50
110
60 4 25 3
60
51 9
28 4 8 8
10

60　821　60

a'

详图1 detail 1

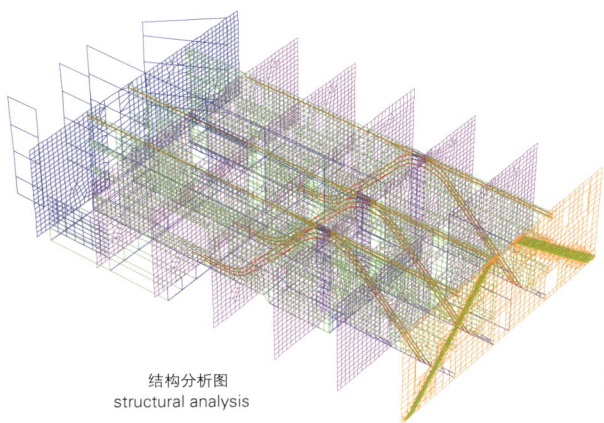

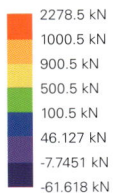

结构分析图
structural analysis

2278.5 kN
1000.5 kN
900.5 kN
500.5 kN
100.5 kN
46.127 kN
-7.7451 kN
-61.618 kN

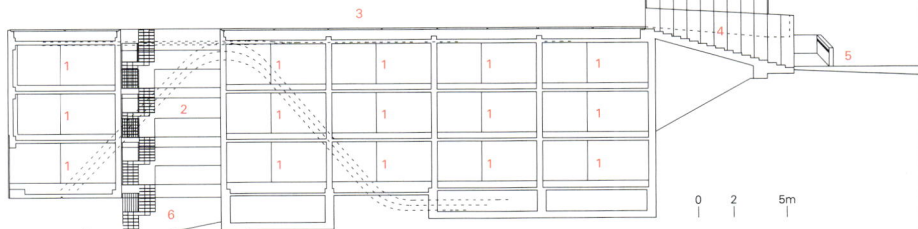

1.私人房间
2.露天庭院1
3.屋顶广场
4.大楼梯
5.入口道路
6.公共空间

1. private room
2. light court 1
3. rooftop plaza
4. grand staircase
5. entrance approach
6. public space

A–A' 剖面图
section A-A'

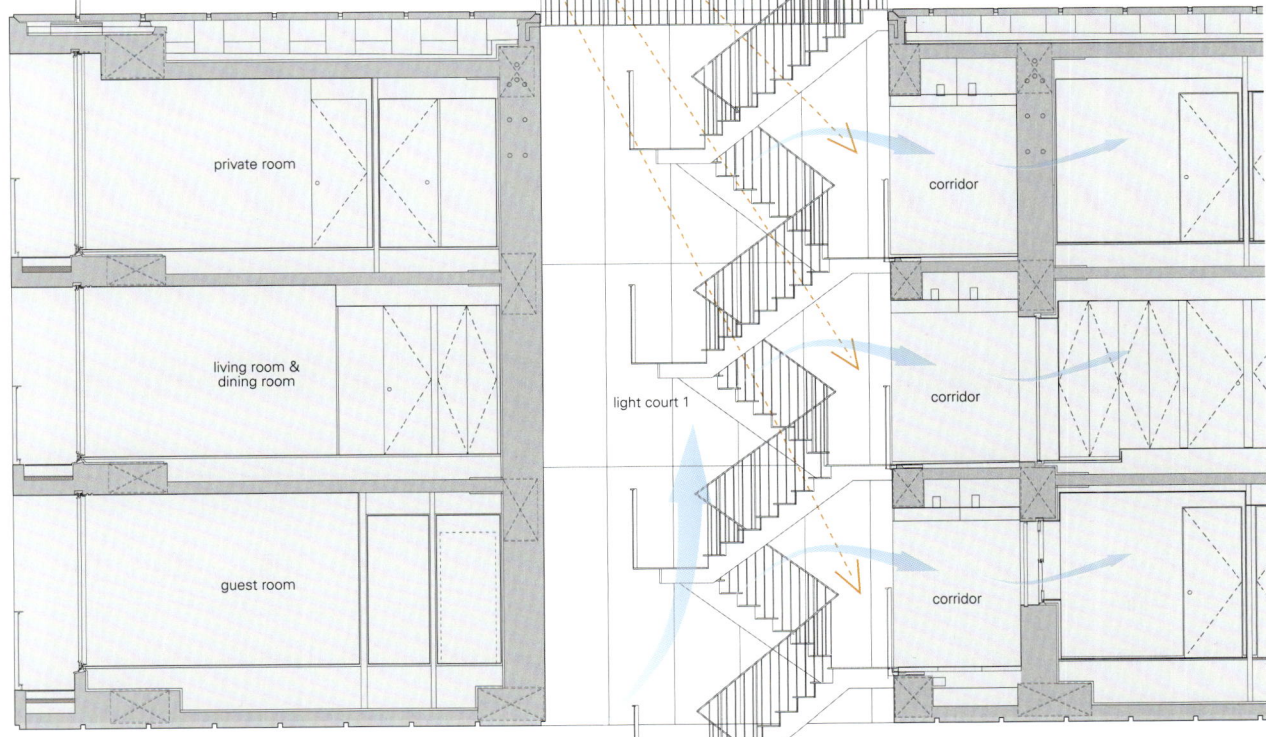

▽4FL=GL +2,600

▽3FL=GL - 700

▽2FL=GL -3,600

▽1FL=GL -6,500

sunlight

rooftop plaza

private room

living room &
dining room

guest room

light court 1

corridor

corridor

corridor

public space

Seto Inland Sea air

0   1   2m

b–b' 剖面详图 detail section b-b'

What we planned was a three-layered structure for habitation, half of it overhanging the cliff. The rooftop with a superb view of Seto Inland Sea has direct access from the north road via the grand stair, and is made open to the public as a rooftop plaza. Under the overhang is another public space with a closed atmosphere that covers people from rain.

Two light courts reminiscent of Cave Dwellings in China are at the same time used as paths to move down to the residential area. They let enough light and breeze in a dreary central hallway and drastically improve lighting and ventilation in each room throughout the building.

The shape of a ship that waits for its launching ceremony is not just a design. It's a figure that true balance of architectural autonomy, and requirements from the city, environment, structure and economy is echoed through a ship, an existence that finds its identity in "balance".

Foundation piles needed enough distance from the cliff edge in order to create a large public space on the roof without stressing weak cliffs. So we adopted cantilever erection method. The adaptation was possible by "continuous structural walls" that function as 3-story-high beams, and "prestressed concrete (PC) cables" that run in themselves curved, tensed according to the stress.

Efficiency of a method to put PC cable in the curved line has already been confirmed theoretically. However the method hasn't virtually been applied yet in the world due to the difficulty in practice, which means, this was the leading-edge technical trial. With this technical breakthrough, it is certain that this new radical structure will be common practice. Therefore, the residential tower with a splendid view serves as a counter weight to stabilize the cantilever building volume.

项目名称：Seto Company Housing / 地点：Fukuyama, Hiroshima
建筑师：Masahiro Harada + Mao Harada
设计团队：Naoto Ishii, Tetsuya Mizukami
总承包商：Shimizu Corporation
客户：Shipbuilding Company
用途：company housing
用地面积：1,934.84m² / 总建筑面积：1,098.38m² / 总楼层面积：3,095.74m² / 楼层规模：8 story
结构：reinforced concrete
室外覆层材料：concrete (form: larch plywood, lauan plywood)
室内覆层材料：concrete (form: arch plywood, lauan plywood), MDF, oak flooring, corten steel (staircase and handrail),
steal sash, aluminum sash (opening)
设计时间：2010.10~2011.8 / 施工时间：2011.10~2013.3 / 竣工时间：2013
摄影师：©Ken'ichi Suzuki (courtesy of the architect)

Brugg镇的阶梯式住宅

Ken Architekten

　　这个充满生机的住宅综合体位于瑞士Brugg小镇边缘南侧斜坡上混合式住宅区内。清水混凝土围墙中一共包含16座独立公寓，且整体的不规则造型展现了建筑设计对山地坡度的大胆迎合。整座建筑周围被草地和沥青覆盖，在建筑的东南角，沥青甚至直接铺进大厅。居民可以搭乘斜坡上的电梯到达建筑的八层生活区；也可以通过电梯井上层的屋顶平台（作为两层楼梯）到达阶梯式公共区或顶楼的操场。总之，电梯和楼梯就像建筑的脊梁，从混凝土外墙方向看清晰可见。在电梯两侧的各层公寓都平行于坡面而建，并总体面向宽敞的屋顶露台。

　　不论是有倾斜顶棚的厚水泥护栏，还是水泥外墙，都可以阻挡外界的噪音，以及路人和邻居的窥视。此外，一层为停车场，从而阻挡了建筑整体，尤其是二层公寓的花园露台（面向街道开放）。每个公寓单元有两个入口：一个大型入口，直接连接公寓后身的电梯；另一个紧挨建筑立面，与楼梯相连接。从楼梯到室外墙体，起居空间沿路都是落地窗设计。如果对公寓进行重新布局，还可以用轻质墙体结构进行改进。

　　入口到浴室之间的空间设有辅助空间，用于将生活区与坡面地下室分隔开来；其中有未供暖的储藏室和洗衣房作为缓冲区。小型浴室和地下室的窗户呈不规则的图案形状，穿插在周围的墙壁上。在临街的墙面和电梯、楼梯侧墙体上也有相似的洞口，以保证日光和新鲜空气可以进入车库和楼梯中。

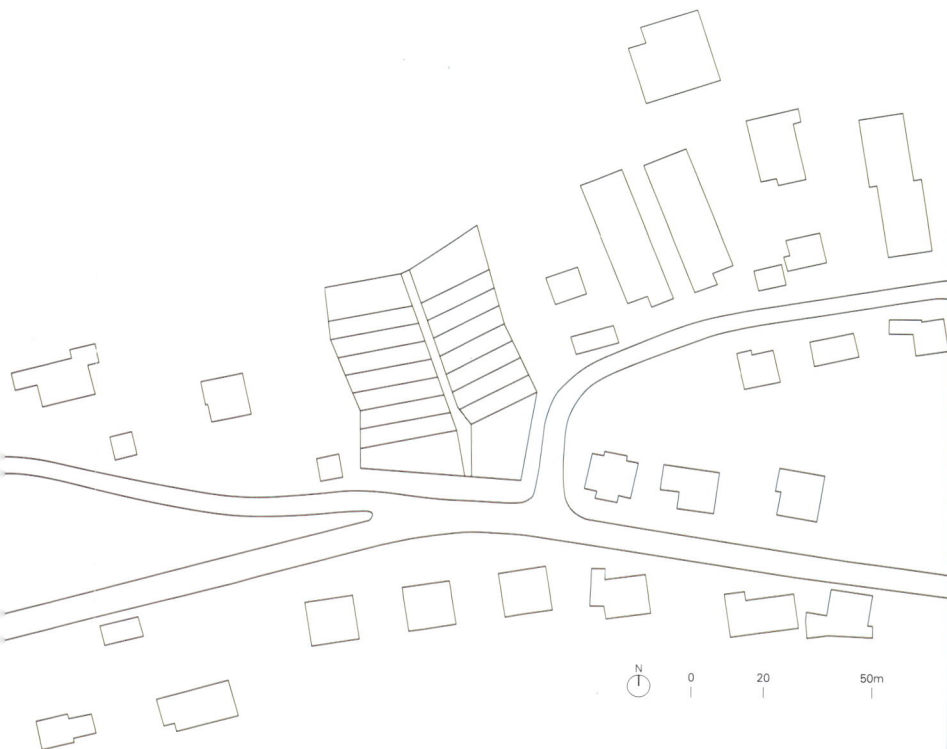

N　　0　　20　　50m

Terraced Housing in Brugg

This robust residential complex is located in heterogeneous residential fabric on a southerly slope on the edge of the Swiss town of Brugg. A perimeter wall of exposed concrete holds together 16 condominium apartments in an irregular form that responds boldly to the topography of the slope. The surfaces surrounding the new structure – meadow and asphalt – abut the building. On the building's southeast corner, the asphalt even flows into the lobby. The sloping elevator situated here serves the eight living levels; the ceiling deck right above the elevator shaft doubles as cascading stair that leads to the communal terrace and playground on the uppermost level. Together, stair and elevator form the spine, which is clearly legible on the exterior. On both sides of it the apartment units are arranged in layers of space that run parallel to the slope and are oriented to the spacious roof terraces. Both the thick concrete parapets with oblique tops and concrete walls shield these outdoor spaces from noise and visual contact with passer-by and neighbours. Moreover, the ground-level parking garage screens the entire ensemble – and in particular, the roof gardens of the first floor flats – from the street.

Each unit has two entrances: a large entrance connected directly to the elevator at the rear of the apartment and, adjacent to the facade, a link to the stairway. From there the living area extends along the completely glazed front to the outer wall. Should separate rooms be desired, lightweight wall construction may be employed.

The zone situated between the entrance and the bathroom contains auxiliary spaces and separates the living spaces from the slope-side basement; unheated storages spaces and laundry rooms serve as a buffer. Small bathroom and basement windows, which are arranged in an irregular pattern, perforate the perimeter walls. Similar openings in the wall facing the street and along the spine allow daylight and fresh air to enter the garage and stairway.

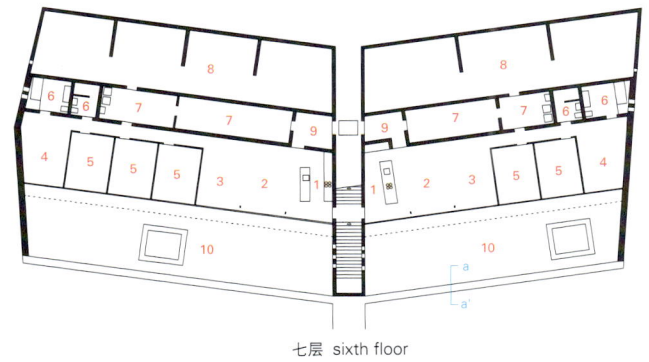

二层 first floor

七层 sixth floor

detail 1

1.厨房　　　　5.卧室　　　　9.入口
2.餐厅　　　　6.浴室　　　　10.露台
3.起居室　　　7.杂物房
4.主卧室　　　8.酒窖

1. kitchen　　　5. bedroom　　　9. entrance
2. dining room　6. bathroom　　10. terrace
3. living room　7. utility room
4. master bedroom　8. cellar

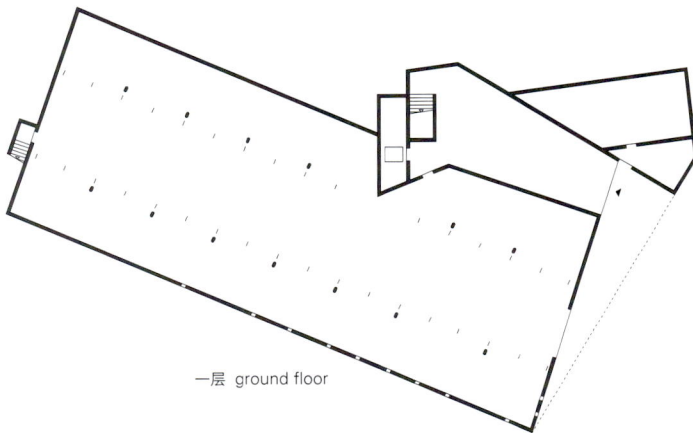

一层 ground floor

三层 second floor

0　　5　　10m

项目名称：Terraced Housing in Brugg
地点：Brugg, Switzerland
建筑师：Ken Architekten
土木工程师：Heyer Kaufmann Partner
客户：Wartmann Immobilien AG
用地面积：4,380m² / 有效楼层面积：2,644m²
设计时间：2008.6 / 施工时间：2013.4
摄影师：©Hannes Henz (courtesy of the architect)

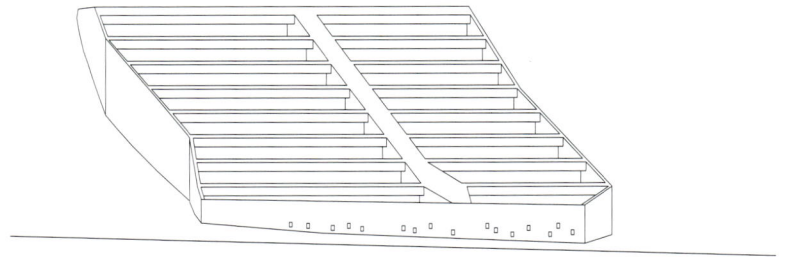

南立面　south elevation

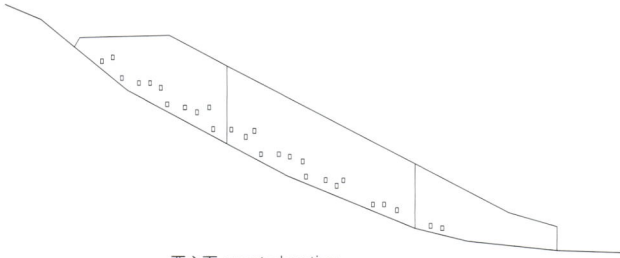

西立面　west elevation

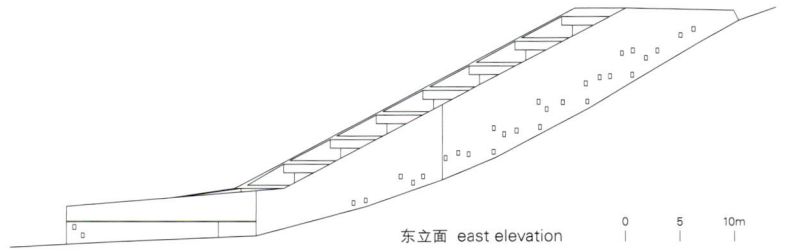

东立面　east elevation

0　5　10m

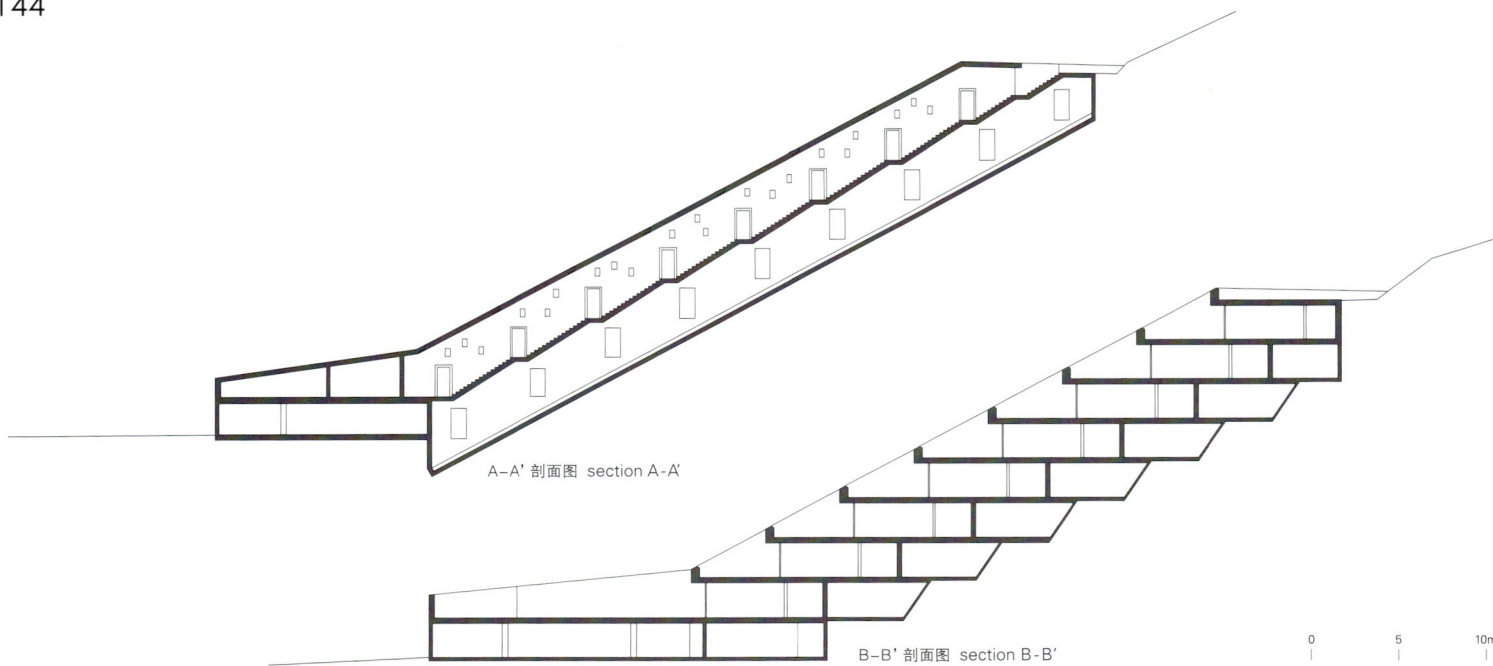

A-A' 剖面图 section A-A'

B-B' 剖面图 section B-B'

0　5　10m

1. cover plate, alu stove-enamelled
2. working joint, vertical with joint tape
3. joint tape
4. ceiling construction
  - cement slab 40mm
  - grit 40mm
  - drainage Enkadrain TP 10mm
  - EP5 WF Ardflam, all-over agglutinated 5mm
  - EGV3 loose, overlapping welded 5mm
  - Swisspor PUR-Alu 140mm
    (0.024W/mK, 30Kg/m³)
  - EVA35, all-over agglutinated 5mm
  - ferroconcrete in slope 240-30mm
  - deposit for sealing 15 x 370 mm
5. throat in wood 20mm x 20mm
6. swelling tape

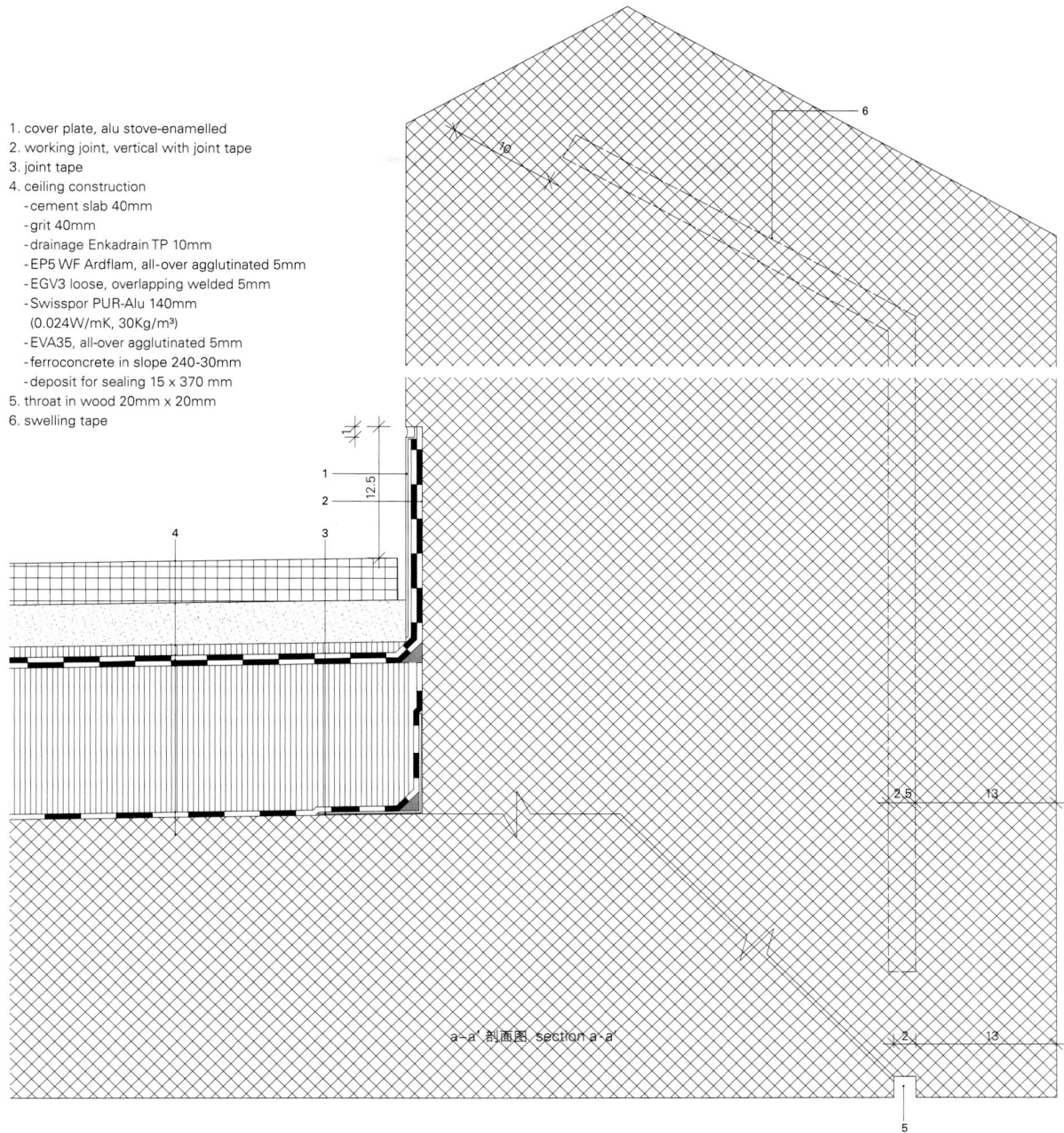

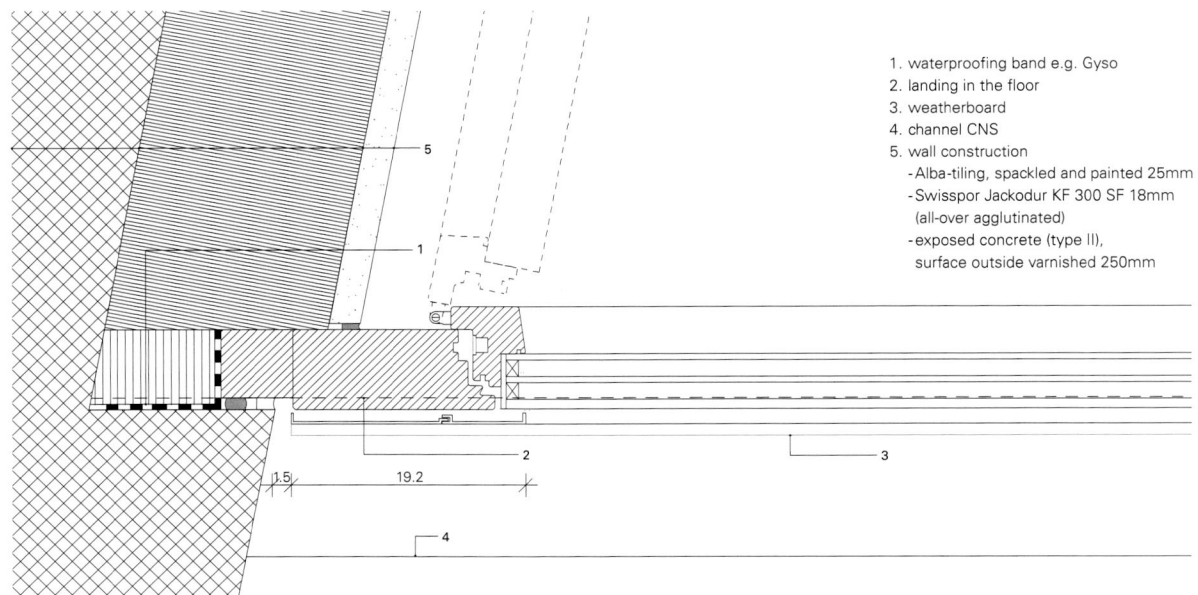

a–a' 剖面图, section a–a'

1. waterproofing band e.g. Gyso
2. landing in the floor
3. weatherboard
4. channel CNS
5. wall construction
  - Alba-tiling, spackled and painted 25mm
  - Swisspor Jackodur KF 300 SF 18mm
    (all-over agglutinated)
  - exposed concrete (type II),
    surface outside varnished 250mm

详图1 detail 1

Tappen住宅

Joliark

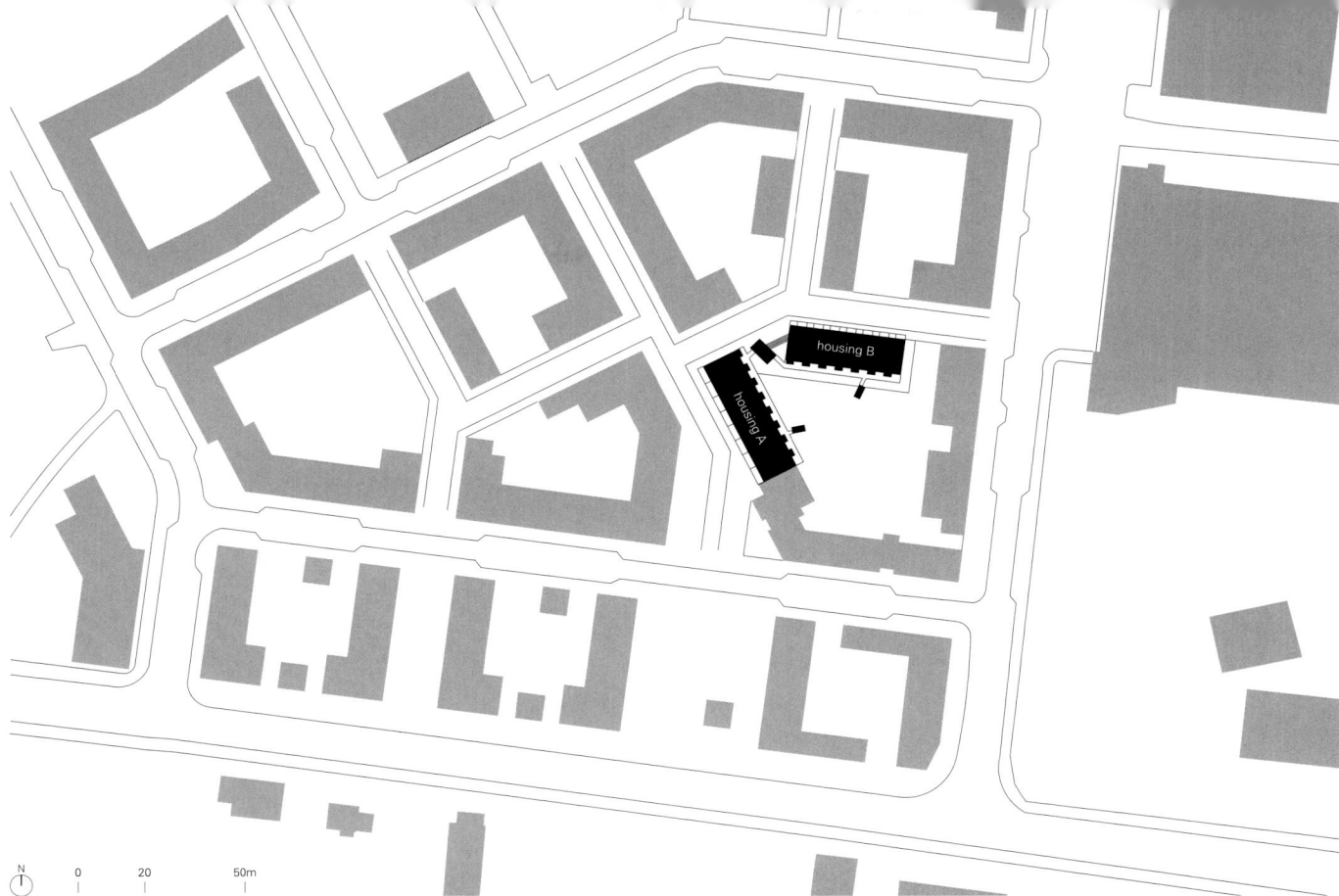

N 0 20 50m

仿佛是在圆年少时的一个梦，建筑师将这个斯德哥尔摩郊外老工业区里的金属公寓大楼设计成了小时候玩偶之家的模样。

此项目在两个体量内建造了44户公寓。保护建筑材料本身的简单纯粹性是这个设计过程中的主要驱动力。木材、混凝土和钢筋之间的紧密组合又维持了建筑的整体感。

整个建筑立面包裹的镀锌钢板就像防弹衣一样，齐平且精细的细节突出了该建筑的形象。而西立面的两层高独立阳台隔间与建筑坚固的外形共同营造出玩偶之家的独特美感。玩偶之家这一隐喻也同时表现出了建筑本身的透明性，阳台定义了公寓的规模，这样路人从外面就能清楚了解建筑的内部结构。

住宅单元经过模块式组合，形成一个拥有23套复式公寓套房的高层体量和一个平层公寓与复式套房共存的体量。两栋金属楼之间是一座朝南的公共花园，且两座楼通过一个玻璃塔楼结构来连接，而塔楼又通过一系列的阳台来与公寓连接。一些通道式露台战略性地设置在花园与厨房之间，从而自然而然地成为人们的社交场所。

公寓里的私人区域用简单直接的方式与公共区域隔开：复式公寓的卧室设在二层，平层公寓的卧室设在北面。内墙的窗户在不损害个人隐私的情况下，满足了公寓中央空间更大的照明透明度需求。

透明性是此建筑的主打理念，以此来模糊室内外的界限。浇筑的肋形墙壁采用裸露混凝土建造，形成阳台的隔间，成为室内墙壁的延续。窗台的天然石板一直铺到地面，而地面下的加热系统则利用石质雕花带分布在木材板边缘齐平的位置。同时，木材也越出室内范围，为阳台添加了一个温暖的布景。

精密的技术也成功解决了建筑的金属外墙与项目中的其他元素无缝连接的难题。屋顶排水系统和排水管道与表面完美结合；镀锌板材在墙角处进行了大角度折叠，将窗框和褶皱都融入到挡水条中。

玻璃塔楼的金属围栏和外墙由多孔板材建成，使其打破了建筑整体给人的材质平滑的印象。塔楼的立面墙板焊接在与混凝土墙相连的框架上，其排列的节奏与窗户的位置相统一。

Tappen Housing

Having nourished an unfulfilled dream of his own doll's house since boyhood, the architect has finally seen it take real shape in this metallic apartment complex on a former industrial site outside Stockholm.

The project comprises 44 dwellings organised within two volumes. A driving force in the design process has been the preservation of a pure and simple material expression. Rigorously composed meetings between wood, concrete and steel ensure the maintenance of a sense of whole.

The facade wraps around the structure like a protective cloak of galvanised steel sheet, an image emphasised by flush precision detailing. The thematic solidity of the project juxtaposed by its western elevation where double height balcony compartments stand creates the explicit doll's house aesthetic.

As a metaphor the doll's house also relates particularly well to the legibility of the buildings, so the internal arrangement manifests itself clearly to the passer-by as the balconies define the scope of each dwelling.

项目名称：Tappen Housing
地点：Mariehäll, Stockholm, Sweden
建筑师：Per Johanson_architect in charge, Stina Johansson, Amanda Hedman
客户：Reinhold Gustafsson Förvaltnings AB
用地面积：2,000m²
总建筑面积：1,400m²
有效楼层面积：5,300m²
施工时间：2014
竣工时间：2015
摄影师：©Note Designstudio (courtesy of the architect)

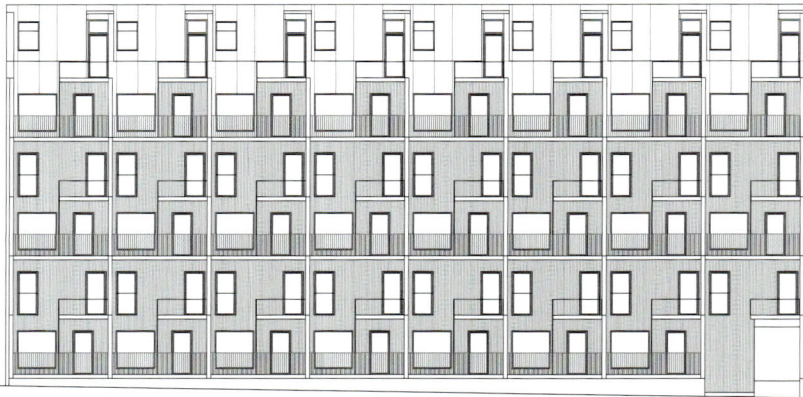

西南立面_A楼
south-west elevation_housing A

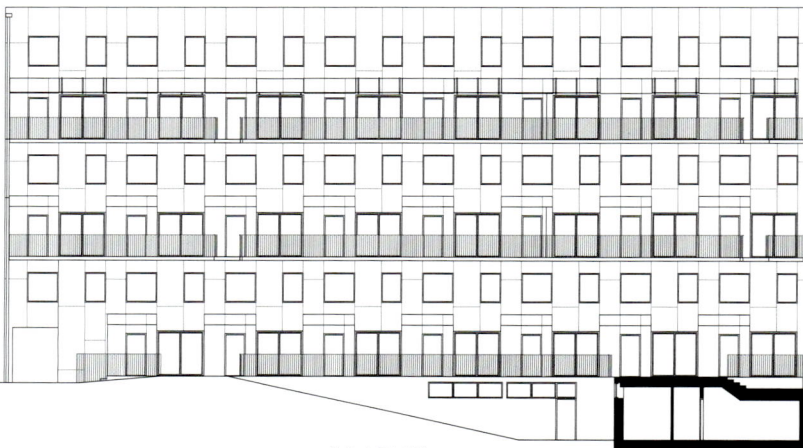

东北立面_A楼
north-east elevation_housing A

北立面_B楼
north elevation_housing B

0　2　5m

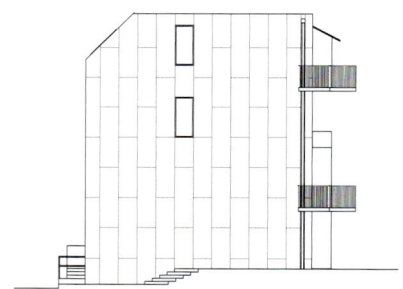

西立面_B楼
west elevation_housing B

西南立面详图_A楼
south-west detail elevation_housing A

- - - shadow joint axis
- - - edge of the panel

25.120 height to the bottom (upper) profile window frame
26.948 height to the bottom (upper) window glass

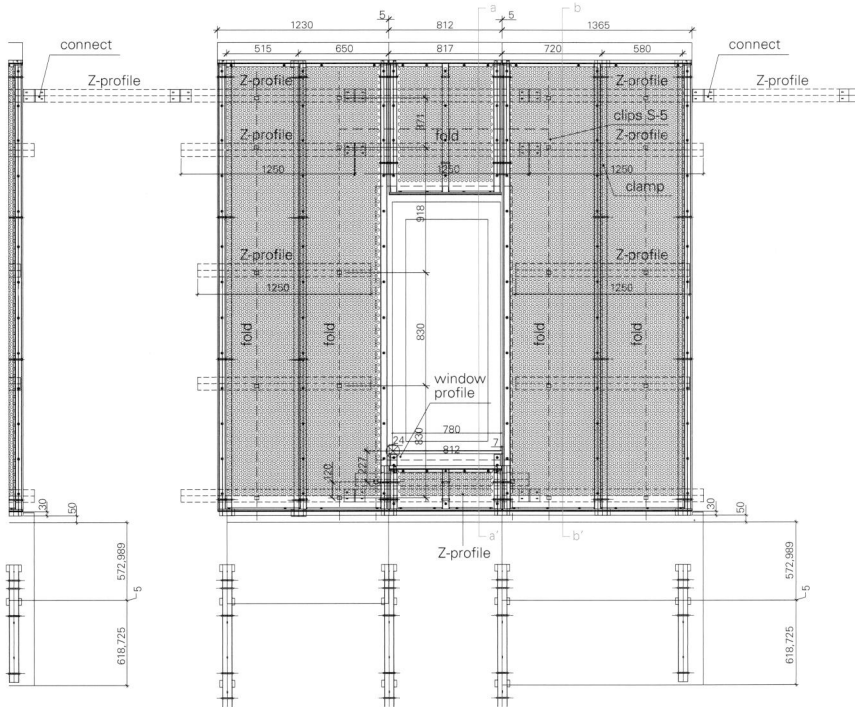

详图1 detail 1

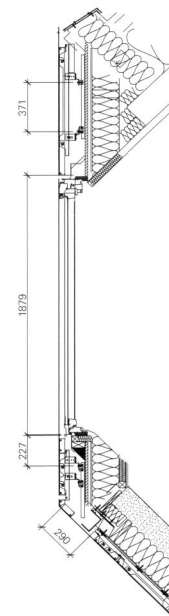

a-a' 剖面图
section a-a'

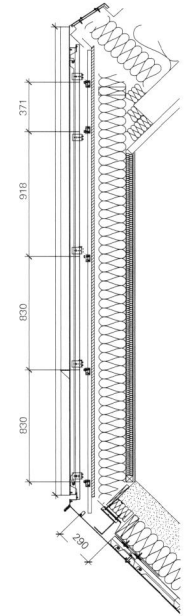

b-b' 剖面图
section b-b'

The units are modularly assembled into a taller volume stacking 23 maisonettes and a shorter one where maisonettes and single storey apartments collaborate. A south facing communal garden separates the two metal clad buildings that are connected by a glazed tower distributing access to the apartments by a system of balconies. Strategically positioned between the garden and the kitchen these access terraces become a natural place for social interaction.

The private zones of the apartments are set apart from the more public areas by straightforward means, collecting the bedrooms upstairs in the maisonettes or facing north in the single storey dwellings. Internal wall openings illuminate central spaces to generate a greater volumetric transparency without compromising the sense of privacy.

Openness is a common theme throughout, blurring the boundaries between inside and outside. The cast fin walls framing the balcony compartments continue internally as an exposed concrete wall. Natural stone slabs in the window ledges spill down on the floor where the heating is distributed via a stone frieze sitting flush with the timber boards. The wood in turn transcends the domestic realm to form a warm backdrop within the balconies.

The seamless character of the snug metal envelope holding the project elements together has been achieved through meticulous technical solutions. Roof gutter and drainpipes have been carefully integrated into the surface, and the galvanised fabric folds sharply over corners, tucks away window frames and pleats into a weather bar.

Perforated sheet make up the metal balustrades and coats of the towers to further enhance the smooth material impression. The facade panels are welded to a framework that connects directly to the concrete wall, their rhythm harmonizing with the spacing of windows.

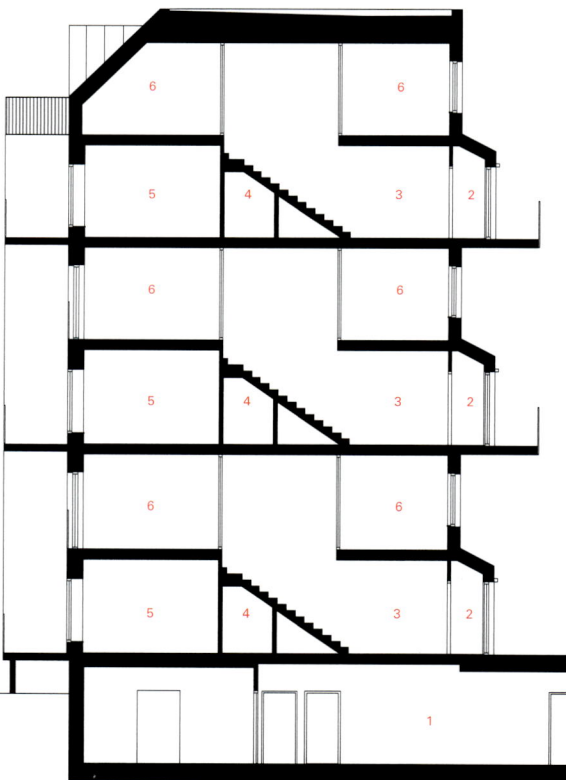

1.停车场
2.入口
3.厨房
4.浴室
5.起居室
6.卧室

1. parking
2. entrance
3. kitchen
4. bathroom
5. living room
6. bedroom

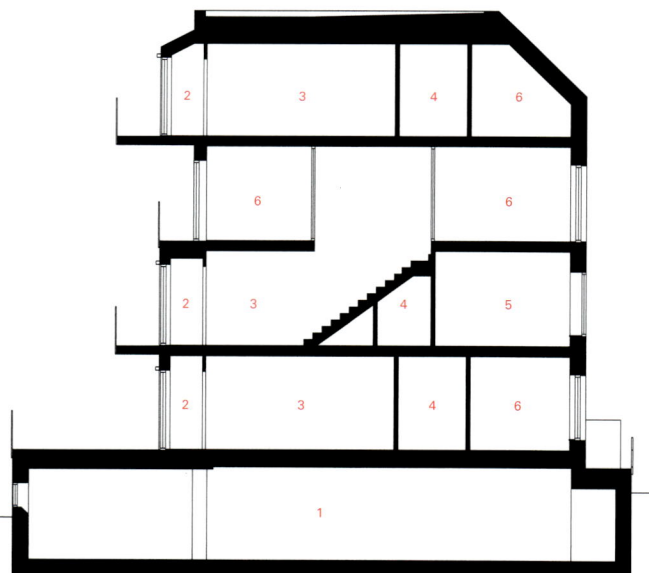

A-A' 剖面图
section A-A'

B-B' 剖面图
section B-B'

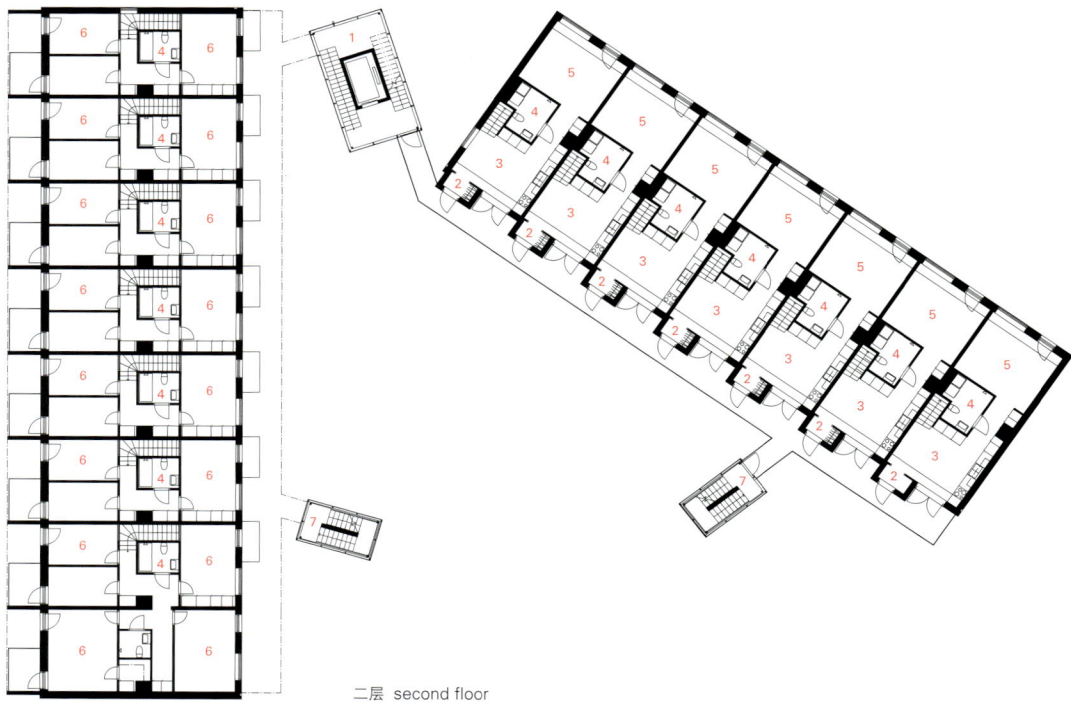

二层 second floor

一层 first floor

1.主楼梯间
2.入口
3.厨房
4.起居室
5.浴室
6.卧室
7.紧急楼梯间

1. main stairwell
2. entrance
3. kitchen
4. bathroom
5. living room
6. bedroom
7. emergency stairwell

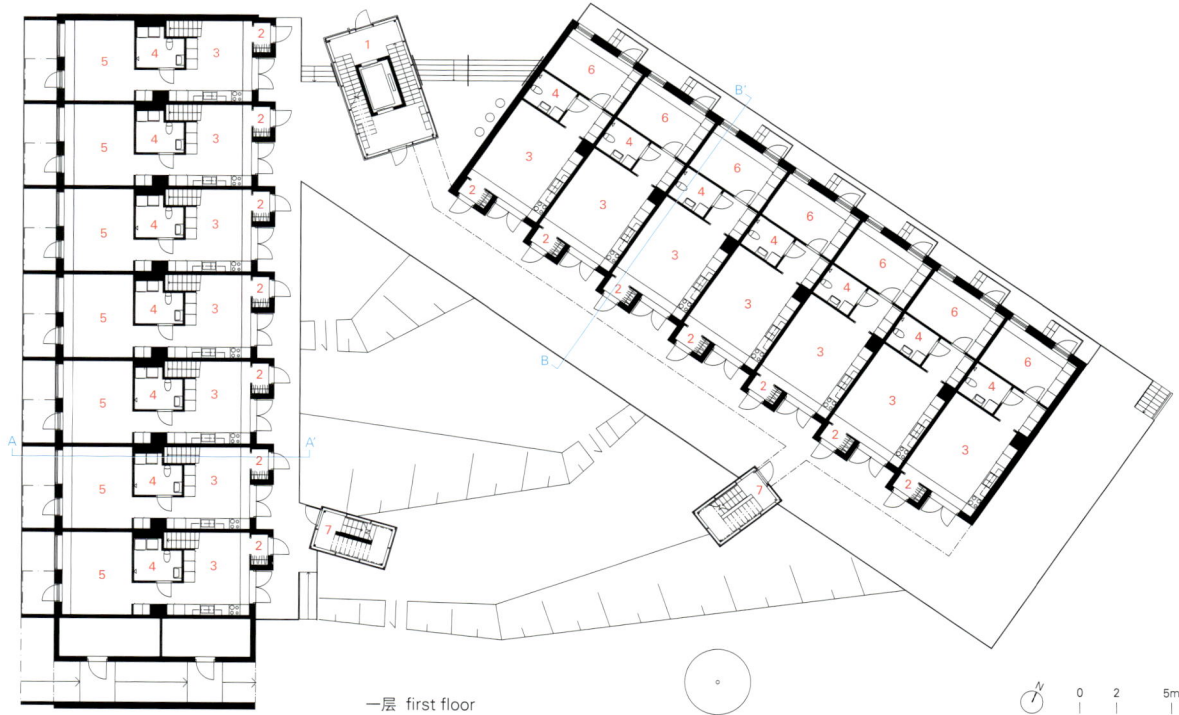

0 2 5m

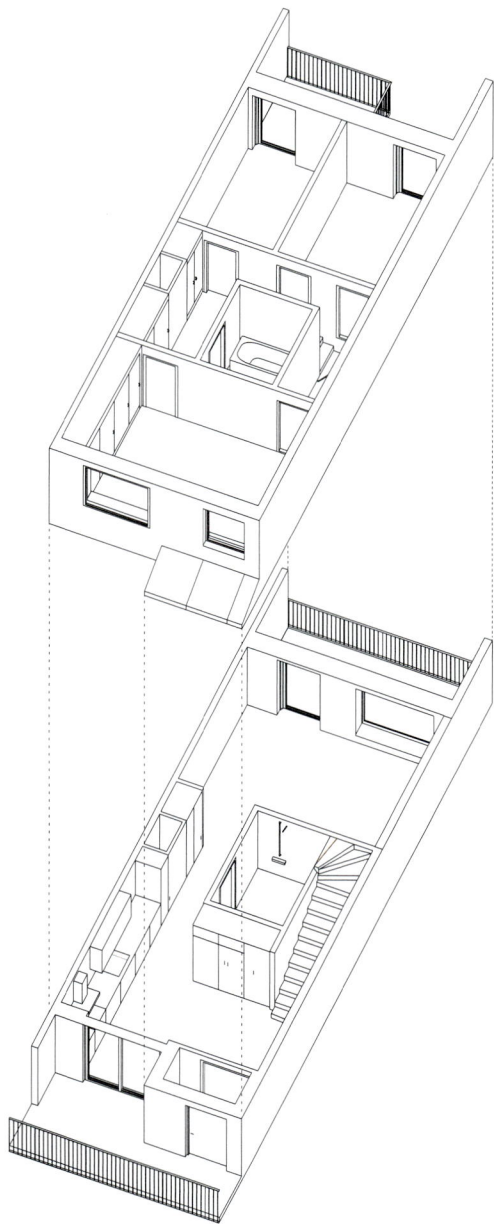

Nantes地区某住宅

Antonini Darmon

Nantes地区建筑的流行风格普遍有了变化后，我们的建筑项目也采用了当地很多独有的元素，并从这些元素的重新诠释中汲取美感。这个项目地处极佳位置，成为了当地的一个标志性聚会建筑。客户的目标是建立一个能够尊重并且回应周围即时环境的社会住宅项目，而这个项目即便从河对岸的Nantes历史古城看过来，也有很高的能见度。项目本身不止有探索和开发现有地形的目标，而且还要成为能代表客户的建筑；如今，"Oiseau des Iles"项目的第一个挑战已经完成，已经成为当地居民心中的显著的地标建筑。

这个项目的功能及其内部衔接的强大之处在于它的双重性，这种双重性存在于地基、单一体量（内设商店，房屋中介）和社区住宿区内。该建筑体块深深嵌在地表，与地面连接并且融入周围环境，也将住宅楼、城市和市民紧紧相连。

社区大楼修长的楼体直指天际，建筑的最高点就像是迎风破浪的船头，简单而坚固。为了迎合城市的线条，建筑立面表皮呈现富有活力的挑窗和栅格结构，它们可在实体、空间以及百叶窗的开合之间转换。

公寓类型有独门式、复式，多数为标准式，每种类型的公寓都有充足的生活空间和一流的室外空间。我们的设计理念也主要体现在这些空间中。一层的花园中设有为居民遮风挡雨的公共空间和露台。塔楼内的室外空间是从露天空间中扩展出来的条状空间。这些空间是封闭的，作为室内洗衣房、酒窖或者其他一些功能房间来使用，它们也能起到过滤的作用，人们可以在室内欣赏外面的景色而不用担心被室外窥探隐私。如果你将这些空间全部打开，则可以欣赏到窗外美景。

混凝土结构简单却微妙，能展现出不一样的室外立面效果。建筑体块覆以木板条。横跨边界的阳台（可作储藏用，预防恶劣天气，保护隐私）被包裹在一层彩色的钢筋防风网格中。网格部分是开放的，部分则是实心的，或有多孔板填充，多孔板可以固定住也可以滑动。坡面屋顶上则采用了太阳能电池板。

Housing in Nantes

After the fashion of the general conversion of the Nantes
block, our architecture exploits the special elements of its
environment and draws its beauty from the reinterpretation
of these. Set in an ideal location, the building becomes a
marker for a place to meet. The client's aim was to develop a
social housing project that would be able to respect and re-
spond to its immediate environment. This project has a great
visibility from Nantes historical city on the other bankside.
In addition to the obvious ambition to explore and improve
existing typologies, the project had to be emblematic for the
client. "Oiseau des Iles" project has clearly overridden its first
challenge, and it's now a powerful landmark for the city and
its people.

The great strength of the program and of its articulation
resides in the duality which exists between the telluric plinth,
a monolith on a pedestrian scale, dedicated to the shops and
intermediary housing, and the community accommodation.
Resolutely grounded, linked to the land and blended into the
region, the block is the link between housing, the city and its
users.

The community building is a slender volume rising towards
the sky where it culminates in a point, a steady prow, proving
to be simple but strong. Echoing the urban thread, the facade

东立面 east elevation

北立面 north elevation

西立面 west elevation

1. concrete balcony
2. folding shutters /
   sliding white lacquered aluminum
3. switch thermal bridges
4. metal railings
5. sliding / swing carpentry
   pre-painted aluminum uw = 1.8
6. reinforced concrete slab 18cm
7. aluminum shutters
8. insulation 14cm
9. reinforced concrete wall 20cm
10. secondary structure facade
11. opaque composite panel

详图1 detail 1

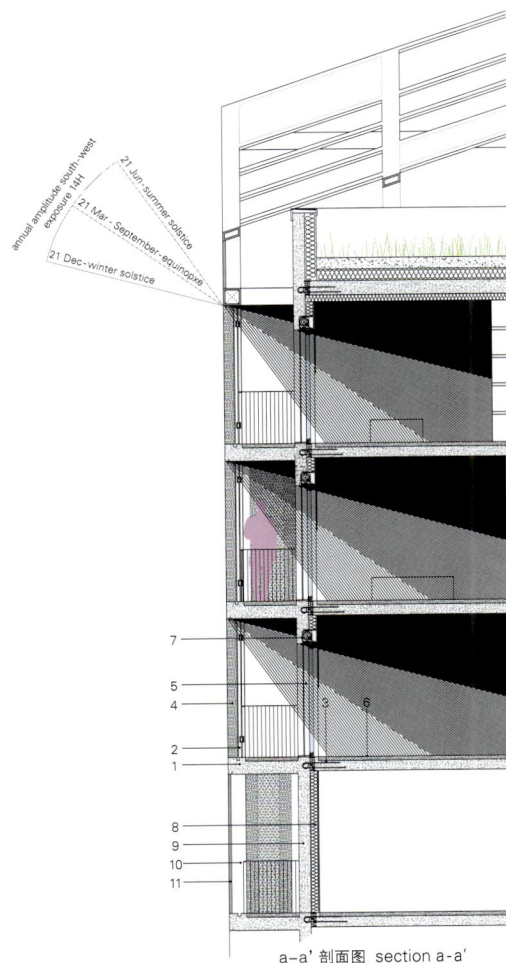
a-a' 剖面图 section a-a'

四层 fourth floor

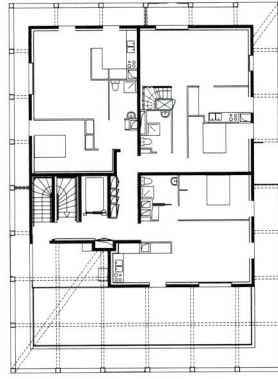

八层 eighth floor

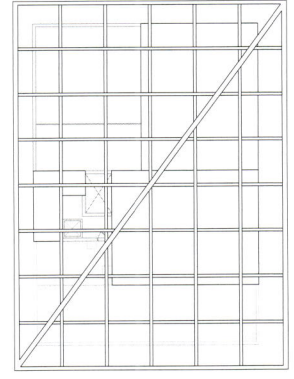

屋顶 roof

三层 third floor

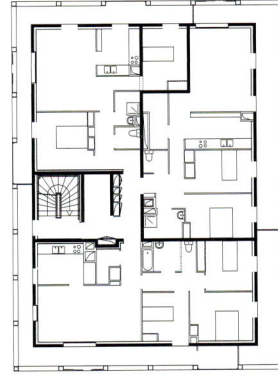

六层 sixth floor

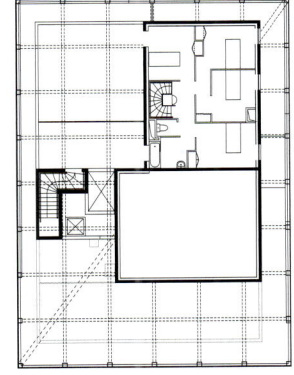

九层 ninth floor

一层 first floor

N 0 2 5m

receives a skin, an almost living moucharaby or lattice work, alternating between solid, void and the movement of the shutters as they are opened and closed.

Offering great outdoor spaces, accommodation is made of individual houses, duplex, and more standard flats, that all having generous angle living spaces and remarkable outdoor spaces. Our reflection was mostly dedicated to these spaces. On the ground floor, garden is made that also buffers with public spaces and patios that protect the inhabitants from the climate. Outdoor spaces within the "tower" are in sling that expands among the exposition, and they can blind in order to

house laundry, cellars, or others. They can filter, see but not be seen. They can open fully to benefit all great views around. Simple but subtle, the structure in concrete receives different teatments of external facade. Block is covered with wooden laths. In the building, full-perimeter balconies (storage, weather protection, privacy...) are enclosed in a square, wind-breaking grid in painted steel. The cells of the grid are partly open, partly screened with solid infill or perforated panels, either fixed or sliding. The slopping roof incorporates solar panels.

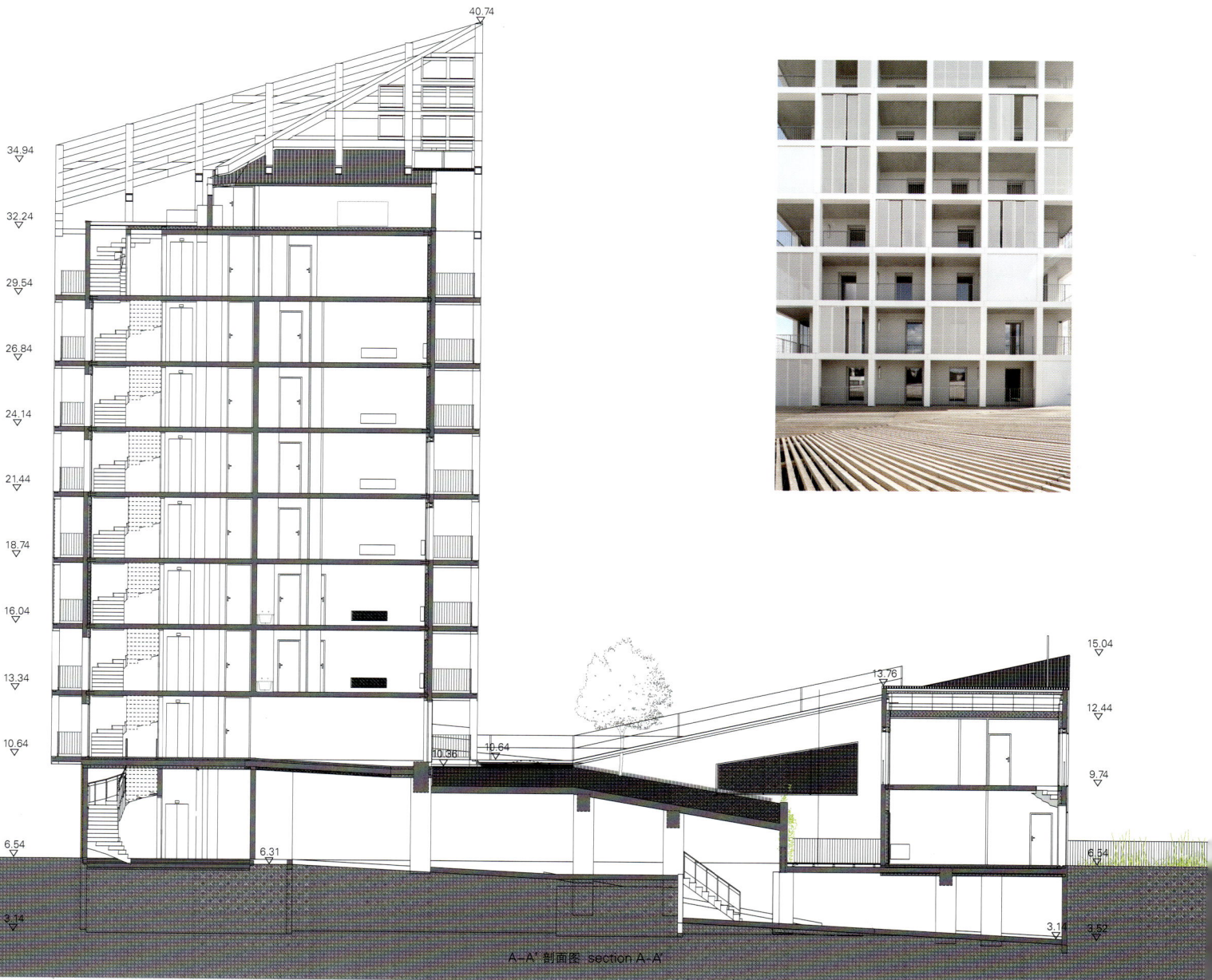

A-A' 剖面图 section A-A'

项目名称：Housing in Nantes
地点：Nantes, France / 建筑师：Antonini Darmon / 项目团队：Laetitia Antonini, Tom Darmon
合作者：Claire Archimbaud _ conception phase, Daliana Vasilache _ construction Phase / 用地面积：2,209m² / 总建筑面积：2,260m² / 有效楼层面积：3,256m²
设计时间：2010 / 竣工时间：2014 / 摄影师：©Alexandre Wasilewski (courtesy of the architect)

b-b' 剖面图
section b-b'

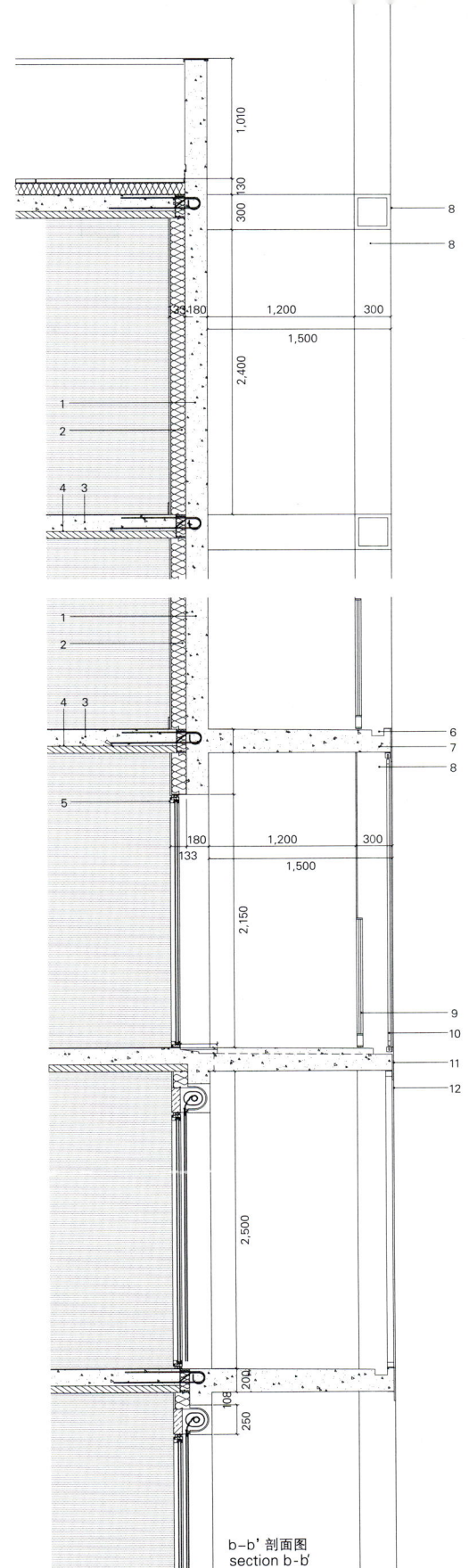

1. masonry wall: 18cm
2. insulation type pregymax 29.5: 12cm + BA13
3. pre concrete slab concrete 20/23cm
4. switch thermal bridges
5. type aluminum carpentry technal soleal home or technically equivalent
6. booking gutter
7. balcony relates concrete thickness 20cm
8. post profile nonstructural H
9. grade body with bars in powder coated aluminum or galvanized steel
10. sliding shutter folding powder coated aluminum or galvanized steel
11. white powder coated steel sheet cladding
12. filling coated aluminum sheet

Grand-Pré 住宅

Luscher Architectes SA

该社区计划提出一种"差异化"的城市结构，从市中心向外逐渐降低密集化程度，最后在城郊形成"空旷"的农耕区。这是一个转型的过程，利用低密度的绿化城市转化和扩建从已建成的环境向"绿植"环境转型。

该建筑利用"镜面"效果和城市之间建立联系，这是一种以小镇，尤其是Grand-Pré路为轴的城市结构进行的换位。它是在分析得出一连串的空间顺序后，形成的一种对空间性、规模、视觉性和人体渗透性的扩展。

除此之外，城市项目还提出了一种混合式建筑，将不同的环境因素凝炼到复杂的场地视觉效果中。

该建筑项目必须实行区域规划的城市理念。

石灰石"建筑"理念的提出改变了建成区域的规模，并且物化了表达效果，使其与小镇的矿物元素建立了联系。

Grand-Pré区的室外空间以由矿物到植物，从公共到私人的过渡为基础。南部区域的一条沿河步行街（修建）会绕回到公共区域中，而私人区域也融入一大片草地中。

Housing of Grand-Pré

The district plan suggests a "differentiated" urban structure, with a gradual decrease in density, starting from the town and all the way to farmland "emptiness". A gradual conversion is formed from a "built" environment to a "planted" one with a green urban transition, expansion or extension of the low density area.

The connection with the town is achieved by a "mirror" effect, a transposition of the structural qualities of the town, particularly of the street-front of the Grand-Pré road. It is an extension of its spatiality, scale, visual and pedestrian permeability, with the subsequent sequencing of spaces revealed in the analysis. In addition, the urban project proposes a hybrid combination, merging the various components of the environment into a condensed vision of the site.

The architectural project had to implement the District Plan urban concept.

The proposed addition of travertine "blocks" changes the scale of the built areas and becomes a materialization expression(materializes) in connection with the mineral aspect of the town.

The outdoor areas of the Grand-Pré district are based on the transition from mineral to green, from public to private.

The southern sector will include a pedestrian path along a river returned(restored) to its former open state, while private plots are integrated into a large meadow.

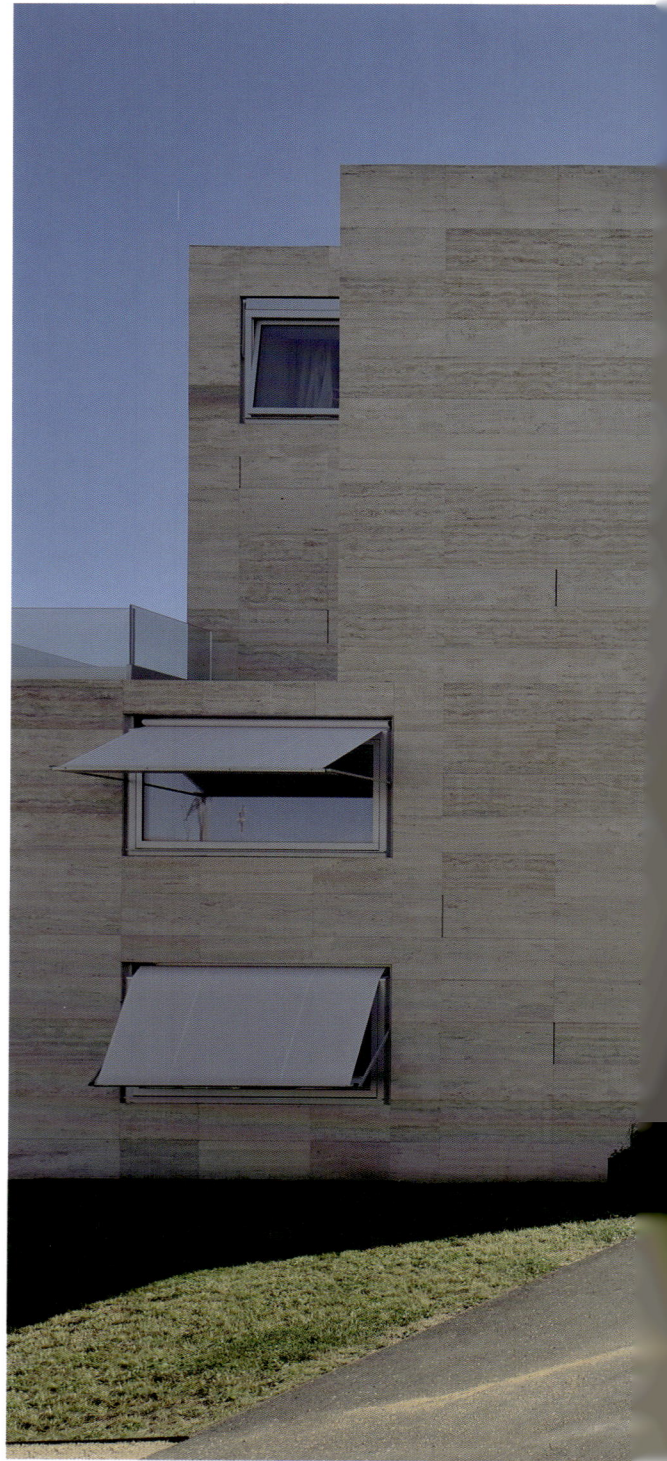

西北立面 north-west elevation

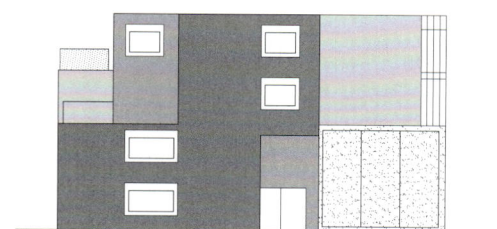

西南立面 south-west elevation

0 2 5m

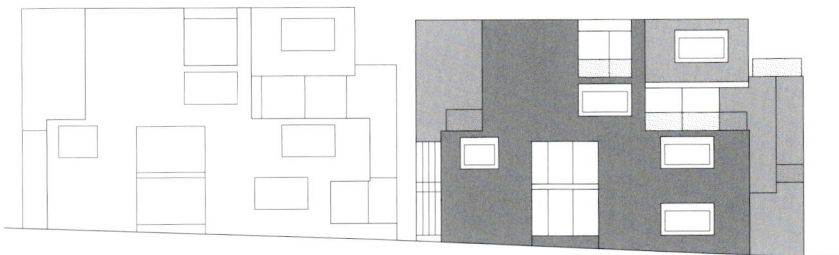

东北立面 north-east elevation

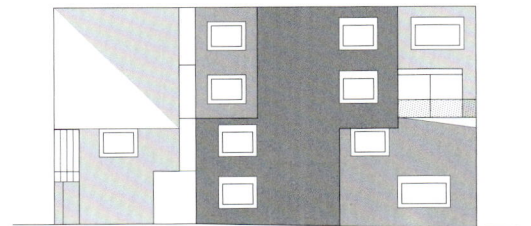

东南立面 south-east elevation

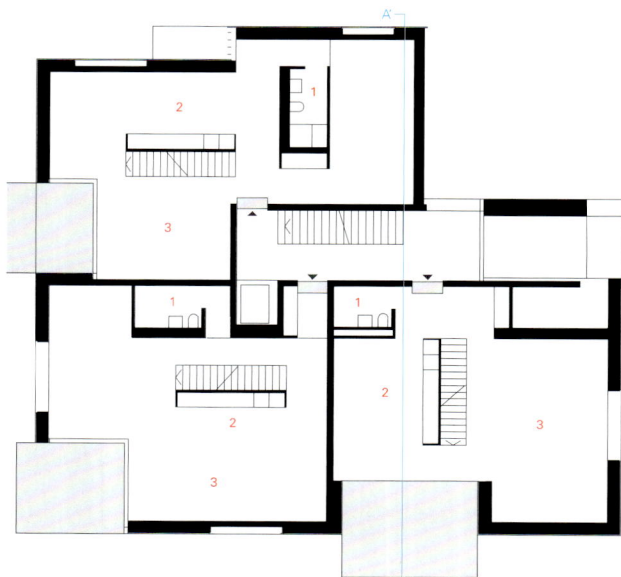

1.浴室 2.厨房 3.起居室
1. bathroom 2. kitchen 3. living room
一层 first floor

1.卧室 2.浴室
1. bedroom 2. bathroom
二层 second floor

1.卧室 2.浴室 3.厨房 4.起居室
1. bedroom 2. bathroom 3. kitchen 4. living room
三层 third floor

1.卧室 2.浴室 3.厨房 4.起居室
1. bedroom 2. bathroom 3. kitchen 4.living room
四层 fourth floor

项目名称：Housing of Grand-Pré
地点：Crans-près-Céligny, Vaud, Switzerland
建筑师：Rodolphe Luscher
项目经理：Mario Da Campo
总建筑面积：6,169m²
设计时间：2008
施工时间：2010
竣工时间：2012
摄影师：©Pierre Boss (courtesy of the architect)

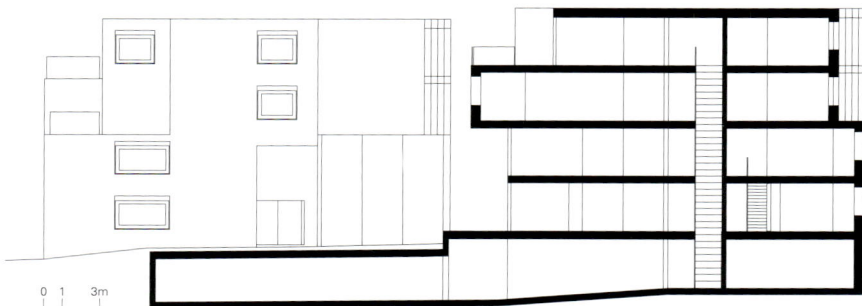

0 1 3m

A-A' 剖面图 section A-A'

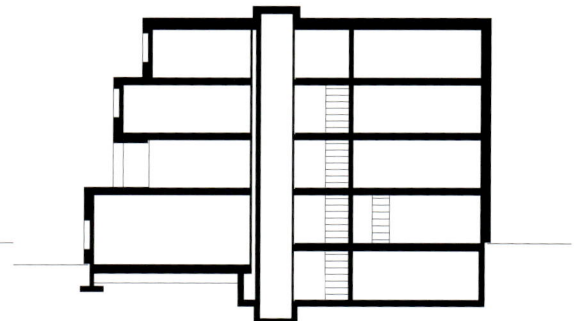

B-B' 剖面图 section B-B'

>>40

OFIS Arhitekti

Was founded by Slovenian architects, Špela Videčnik and Rok Oman in 1997. They graduated from the School of Architecture of Ljubljana in 1997 and received M.Arch from AA School, London in 2000. Some of their works were nominated at the Mies van der Rohe Awards 2006 and 2008. Have been teaching at the Graduate School of Design, Harvard University since 2012. They always start design with the search for a critical issue about the program, site or the client. Their design is far from surpassing, confronting, ignoring or disobeying the rules and limitations of each project. They plunge right in the middle of them and obey the 'law' literally instead. They sometimes attempt to exaggerate it if needed. In their work, restrictions become opportunities for an architectural system.

>>170

Luscher Architectes SA

Rodolphe Luscher was born in Zurich, Switzerland in 1941 and began his career while attending the Zurich School of Art. Having collaborated with Professor Alberto Camenzind in 1961 on the design of the Lausanne National Exhibition center, he left for Norway where he worked with the architect Haakon Mijielva. Subsequently, he forged relationships with Christian Norberg-Schultz and Sverre Fehn. Founded his architectural practice in Lausanne in 1970 and has taught at the University of Geneva, Federal Polytechnic School of Zurich and Lausanne. Was president of Europan Suisse for over 25 years and vice-president of Europan Europe from 1997 to 2002. Established his own office in 1997.

>>56

Baumhauer

Philipp Baumhauer was born in Munich in 1974 and he moved to Berlin in 1995 to study architecture at TU Berlin. After earning diploma, he worked for several architectural offices in Berlin and New York. In 2010, founded his own architectural office and today leads a team of architects working on different projects in different cities around Europe. For Baumhauer, the joy of discovering new questions and their answers is the main focus.
Baumhauer, is also constantly looking for new challenges through collaborations. This allows them to think across multiple disciplines and work, resulting in a series of architectural products that range in scale from abstract, seemingly purposeless objects and images all the way up to buildings and urban planning.

>>108

Lorcan O'Herlihy Architects

Is committed to engage the complexities of contemporary society through architecture. Approach their work with ruthless optimism, dedicated to a conviction that architecture can awaken people and enrich communities. Their process is collaborative and iterative. They understand that their architecture is accomplished by and for people. Since 1990, LOHA has built over 75 projects across three continents. Their work ranges in typology from institutional buildings to bus shelters, and from large-scale developments to single-family homes. Have been published in over 20 countries and recognized with over 100 awards, including the 2010 AIA Los Angeles Firm of the Year.

>>136

Ken Architekten

Lorenz Peter[middle] Jürg Kaiser[left] and Martin Schwager[right] were born in 1968 and received Bachelor of Science in Architecture from the ETH Zurich.
Lorenz Peter has worked as Design assistant at ETH Zurich for 2 years before founding his own office in Zurich in 2000. Has been working as a Partner of Ken Architekten since 2003. Jürg Kaiser studied Structural draughtsman at the Hirschthal and founded Ken Architekten in 1995 with Martin Schwager. Martin Schwager studied Structural draughtsman at the Dietikon and founded Ken Architekten in 1995 with Jürg Kaiser. Is a Technical consultant for site constructions in Brugg since 2014.

>>170

Antonini Darmon

Architecte D.PL.G, Laetitia Antonini(1975)[right] and Tom Darmon(1978)[left] are graduates of Architecture School Paris-Val-de-Seine. Founded the agency Antonini + Darmon Architects in Paris in 2006. Are winners of the 40 emerging architects European under 40(40 under 40).
Their creative system was initiated from the partnership of two complementary yet different personalities. They aim to propose an original architectural interpretation, where each project follows a path having its own identity and dynamics. Their architecture is contextual: functionality, rigor and sobriety are combined in order to transform the project into an object of beauty. They are especially more dedicated to the new architectural and urban stakes involved with sustainable development. Their main approach is inspired by the will to create intimate liv-

>>122

Mount Fuji Architects Studio

Was established in 2004 by two Japanese architects, Masahiro Harada[right] and Mao Harada[left].
Masahiro Harada received a M.Arch from the Shibaura Institute of Technology in 1997 and worked at Kengo Kuma & Associates and Arata Isozaki & Associates. Also has worked with Jose Antonio Martinez Lapena and Elias Torres (Barcelona, Spain) as National Fellowship for Artist in 2001 and 2002. Has taught at the University of Keio COE in 2007. Is currently teaching at the University of Tokyo and his alma mater.
Mao Harada received a B.Arch from the Shibaura Institute of Technology in 1999 and has worked at Editorial office of Workshop for Architecture and Urbanism. Has been teaching at the Tohoku University since 2013.

Diego Terna

Received a degree in architecture from the Politecnico di Milano and has worked for Stefano Boeri and Italo Rota. Has been working as critic and collaborating with several international magazines and webzines as editor of architecture sections. In 2012, founded an architectural office, Quinzii Terna together with his partner Chiara Quinzii. Currently is an assistant professor of Politecnico di Milano and runs his personal blog L'architettura immaginata (diegoterna.wordpress.com).

Isabel Potworowski

Has graduated from TU Delft with a Master in Architecture, and currently works for Barcode Architects in Rotterdam. During her graduate studies, she was a member of the editorial committee and wrote several articles for the independent student journal "Pantheon". Originally from Canada, completed her Bachelor in Architecture at McGill University in Montreal, where she was awarded the Louis Robertson book prize for the highest grade. Besides, studied for one semester at the Politecnico di Milano. Has worked at ONPA Architects and Manasc Isaac Architects, both in Edmonton, Canada.

>>68

Omer Arbel

Omer Arbel was born in Jerusalem in 1976 and afterwards moved to Vancouver. Received Bachelors of Environmental Science in 1997 and Professional Bachelors of Architecture in 2000 from the University of Waterloo, School of Architecture. Has worked at the Patkau Architects, Busby+Associates, Bocci Design and Manufacturing in Vancouver before establishing Omer Arbel Office in 2005. Based in Vancouver and Berlin, he explores the intrinsic mechanical, physical, and chemical qualities of materials as fundamental departure of work. His interdisciplinary practice spans multiple scales and cultural-economic contexts to include building design, industrial design, materials research and high craft manufacturing. He was an instructor of the Parsons Scool of Design and of the University of British Columbia, School of Architecture.

>>82

Schemata Architects
Jo Nagasaka was born in Osaka, Japan and graduated from the Department of Architecture, Tokyo University of the Arts. Established Schemata Architects in 1998 right after his graduation. Currently he has an office in Aoyama, Tokyo. Has an extensive experience in a wide range of expertise from furniture to architecture. His design approach is always based on 1:1 scale, regardless of what size he deals with. He works extensively in Japan and abroad, while expanding his design activity in various fields.

>>148

Joliark
Is an architectural office based in Stockholm, Sweden, mainly practicing in the field of residential, office and industrial architecture. Is owned by Per Johanson[above], Helen Johansson, Hans Linnman, Magnus Pörner and Cornelia Thelander. The practice was founded in 1972 following an awarded first prize in a Nordic architectural competition for the design of Vanda Centrum outside Helsinki.

>>92

atelier tho.A
Was founded in 2014 with the first project FA House, shortlisted in World Architecture Festival(WAF) 2015. This is a small studio led by two principals, Pham Nhan Tho(1987) and Pham Phuong Thoa(1990). Both of them had graduated from Ho Chi Minh University of Architecture in 2010 and 2013. They had worked for Vo Trong Nghia and A21 Studio before creating their own studio. The head office of atelier tho.A is in Lam Dong province, and the second office is in Ho Chi Minh city.

图书在版编目(CIP)数据

时间 : 空间记忆 : 汉英对照 / 韩国C3出版公社编 ;
于风军等译. — 大连 : 大连理工大学出版社，2016.9
（C3建筑立场系列丛书）
书名原文: C3:Time: Memory in Space
ISBN 978-7-5685-0546-8

Ⅰ．①时… Ⅱ．①韩… ②于… Ⅲ．①建筑设计—汉
、英 Ⅳ．①TU2

中国版本图书馆CIP数据核字(2016)第199867号

出版发行：大连理工大学出版社
　　　　　（地址：大连市软件园路80号　　邮编：116023）
印　　　刷：上海锦良印刷厂
幅面尺寸：225mm×300mm
印　　张：11.5
出版时间：2016年9月第1版
印刷时间：2016年9月第1次印刷
出 版 人：金英伟
统　　筹：房　磊
责任编辑：许建宁
封面设计：王志峰
责任校对：高　文
书　　号：978-7-5685-0546-8
定　　价：228.00元

发　行：0411-84708842
传　真：0411-84701466
E-mail：12282980@qq.com
URL：http://www.dutp.cn